6.13-19

Cable Television:
Developing Community Services

Rand Cable Television Series
Walter S. Baer, *Series Editor*

•

Cable Television:
A Handbook for Decisionmaking
Walter S. Baer

Cable Television:
Franchising Considerations
Walter S. Baer, Michael Botein,
Leland L. Johnson, Carl Pilnick,
Monroe E. Price, Robert K. Yin

Cable Television:
A Guide to Federal Regulations
Steven R. Rivkin

Cable Television:
Developing Community Services
Polly Carpenter-Huffman,
Richard C. Kletter,
Robert K. Yin

CABLE TELEVISION:
Developing Community Services

Polly Carpenter-Huffman
Richard C. Kletter
Robert K. Yin

CR

Crane, Russak & Company, Inc.

NEW YORK

Cable Television:
Developing Community Services

Published in the United States by
Crane, Russak & Company, Inc.
347 Madison Avenue
New York, N.Y. 10007

ISBN 0-8448-0260-3

LC 73-91375

Printed in the United States of America

Preface

THIS book is the fourth of four volumes that present the results from a Rand Corporation study of cable television. The study was supported by a grant from the National Science Foundation to the Rand Communications Policy Program. A grant from the John and Mary R. Markle Foundation also aided completion of this volume.

Rand began its research on cable television issues in 1969, under grants from The Ford Foundation and The John and Mary R. Markle Foundation. The central interest at that time was federal regulatory policy, still in its formative stages. Rand published more than a dozen reports related to that subject over the next three years. This phase of Rand's concern ended in February 1972 when the Federal Communications Commission issued its *Cable Television Report and Order*.

The *Report and Order* marked the end of a virtual freeze on cable development in the major metropolitan areas that had persisted since 1966. It asserted the FCC's authority to regulate cable development, laid down a number of firm requirements and restrictions, and at the same time permitted considerable latitude to communities in drawing up the terms of their franchises. It expressly encouraged communities to innovate, while reserving the authority to approve or disapprove many of their proposed actions.

The major decisions to be made next, and therefore the major focus of new cable research, will be on the local level. These decisions will be crucially important because cable television is no longer a modest technique for improving rural television reception. It is on the brink of turning into a genuine urban communication system, with profound implications for our entire society. Most important, cable systems in the major markets are yet to be built, and many cities feel great pressure to begin issuing franchises. The decisions shortly to be made will reverberate through the 1980s.

Aware of the importance of these events, the National Science Foundation asked Rand in December 1971 to compile a cable handbook for local decisionmaking. The handbook, Volume I in this series, presents basic information about cable television and outlines the political, social, economic, legal, and technological issues a community will face. This book (Volume IV) explores the uses of a cable system for education, local government services, and public access to television. Other volumes discuss cable technology and issues of franchising (Volume II) and the federal regulations that apply to cable (Volume III).

The entire series is addressed to local government officials, educators, community group members, and other people concerned with the development of cable television in their communities. It also is intended as text and reference material for college and university classes in communications.

The study director, Walter S. Baer, served as editor for the series. Contributors to this volume include Rand research analyst Polly Carpenter-Huffman, Rand research psychologist Robert K. Yin, and Richard C. Kletter, video and film consultant to the Rand Corporation and other organizations. The assistance of Vicky Agee of the Cable Television Information Center, Washington, D.C. in locating materials for Chapter 2 and of Grace Polk of Rand for the initial case study of WNYC-TV is greatly appreciated. The authors also acknowledge the helpful criticisms on chapter drafts by Rudy Bretz, Nathaniel Feldman, Stanley Gerandasy, Jerry Glasshagel, Peter Greenwood, Leland Johnson, Henry McCarty, Milbrey McLaughlin, Frank Norwood and Elinor Richardson.

The views expressed in this book are those of the authors and do not necessarily reflect the opinions or policies of the National Science Foundation or the John and Mary R. Markle Foundation.

July, 1973

Walter S. Baer
Santa Monica, California

CONTENTS

Part I

Making Public Access Effective
Richard C. Kletter

I. THE PROMISE OF PUBLIC ACCESS

Public access is one of cable television's most significant prospects. With programming created by local citizens for local citizens, and transmitted on cable channels dedicated for that specific purpose, television may finally discover local issues and culture. It may become commonplace to see ordinary people videotaping programs in their own neighborhoods, bringing the problems and pleasures of local life to the attention of their communities.

This phenomenon could restore to the television screen some qualities that have nearly been refined out of it: spontaneity; originality; controversy; realism; even attractive amateurishness. But beyond the temporary pleasure of seeing friends' faces and one's own program on the screen, public access cable can offer a forum for ongoing groups active in meeting concrete community needs. Whether they be neighborhood artists or community organizers, these groups are not interested in television for its own sake, but in reaching people with ideas and information. Their programs offer a contrast to standard prepackaged TV fare, rendered soothing and bland for national mass consumption. The challenge of access is to bring these people to television—to present their ideas with integrity and in a way that will interest audiences accustomed to a more polished format.

A word about terminology: "Public access" specifically refers to the minimum of one channel the FCC now requires for that purpose on each new cable system in the 100 largest television markets, to be available on a first-come first-served nondiscriminatory basis.[1] Of course, the public will have some other means of access to cable. It can have access, at a price, to the cable operator's leased channels (when they become available); and the one channel each to be set aside for education and local government can also be regarded as "public." But these are different matters, as discussed below.

The FCC public access rules are interim, pending a review of experiences. If access is to prove viable, present difficulties in bringing new voices to cable and helping them use it effectively must be solved.[2]

Effective public access is difficult to bring about. It is simply not enough to announce that a channel is available to the public. Many citizens will neither

---

[1] The FCC rules are summarized in Walter S. Baer, *Cable Television: A Handbook for Decisionmaking*, Crane, Russak & Co., New York, 1974.

[2] Which at the outset must resolve questions of cable-system control and ownership—issues of particular importance to revenue-starved community groups. For a full discussion of ownership alternatives, see ibid., Chapter 4.

understand nor be interested in public access; others will be afraid of it. Most people regard television production as a highly technical process in which average people cannot possibly participate. For access to serve the entire community, then, strong psychological barriers must be overcome. It will be easy to attract budding TV producers and other users interested in television itself; the challenge is to reach people working on social problems and to demonstrate that access can aid them in their work.

Even those who quickly recognize the potential and step forward will need help to use the equipment and channel time effectively. Organization, training, and funding will be needed from the start.

These are the problems this report addresses. To illustrate possible solutions, I have gathered experiences from a variety of community television projects. These accounts, coupled with analysis of the community's resources, should provide enough information to start interested groups on their own television projects.

The report also describes cable television and the current FCC rules as they relate to access. It emphasizes the importance of community involvement in the franchising process. It furnishes production information, including a discussion of the relative costs and merits of different kinds of equipment. It gives advice on program style and content. Finally, the report offers ideas on financing and publicizing your programs. The Appendixes list people, institutions, and reference materials that might be useful.

The development stage of a new technology is the critical period for bringing public influence to bear. That period has now arrived for cable television, and the public will have to inform itself thoroughly and quickly about cable issues if this technology is to serve social needs. Once institutions or industries have become firmly established, as we learned with radio and TV, it is extremely difficult to reorient them. Cable may well be our last chance to shape an important communication medium to our own desires. Timely analysis is vital. Without it, narrow interests can dazzle the uninitiated with the magic of the technology and cow them with expertise. The public interest can be sold out for a few glittering trinkets. But there must come a time for laying investigation aside; analysis must finally lead to action.

It is ironic that the impending explosion of a visual culture has generated such quantities of print as a prelude. Informative papers and useful studies have become an industry almost as substantial as the media they treat. As I begin producing my own quota of information tonnage, I must urge that less be written and more be done, that more video experiments be supported in communities throughout the country. Otherwise, new media forms will serve the same narrow interests as their predecessors.

LOCAL PROGRAM SOURCES FOR CABLE TELEVISION

Cable television began in areas with inadequate broadcast reception and few channels to choose from. In isolated rural communities, distant suburbs, or hilly cities, cable's master antenna (usually atop a nearby hill) picked signals from the air and delivered them by coaxial cable to subscribers' homes for a monthly fee. Cable added a few channels, eliminated ghosting and other reception problems, and often delivered a better picture than that received in normally "good" reception areas.

After twenty years of limited reception service, cable is coming of age. Cable expands channel capacity well beyond the broadcast signals available even in the largest markets, such as New York or Los Angeles. Within five years, all cable systems in the top 100 television markets will have a minimum of 20 channels; a few systems with a potential for 40 channels have already been constructed.

Cable signals come from: (1) all the broadcast stations in a given reception area; (2) broadcast signals imported from other reception areas; and (3) nonbroadcast signals originating locally, from studios of the cable system, or from a school or other institution.

Recognizing "the need for more outlets for community expression ... and ... cable television's capacity to provide an abundance of channels,"[3] the FCC now requires cable systems in the top 100 markets to provide one nonbroadcast channel for each over-the-air signal carried on the cable. Although most operators believe that subscribers are currently attracted mostly by improved broadcast services, many of them believe that nonbroadcast services will prove to be cable's most lucrative attractions.

These nonbroadcast signals give cable its main chance to be different. Three kinds of channels are available for a wide variety of programming: leased channels, local origination channels, and public access channels.

Leased Channels

The cable operator has the opportunity to lease time-slots or an entire channel to a programmer who then takes responsibility for content, scheduling, and advertising. Leased channels can be used in two general ways: (1) for sponsored programs available to all cable subscribers—for example, auto care information from local car dealers; and (2) for programs available only to subscribers paying for that specific service—that is, pay television. Although little channel leasing has occurred in the past, many observers of the cable industry expect it to be important in the future. New cable systems in the top 100 markets must have channel capacity to lease, according to the FCC rules.

Local Origination

Local origination means programs either produced or distributed by the cable operator. The ones he distributes might include old films, free government and industry films, "how to" tapes on golf with a famous golfer, and potentially anything on film, videotape, and soon, videocassettes. The programs he produces might include local sports, local news and public affairs, children's programming, and talk shows with local citizens and politicians. Advertising is permitted by the FCC at natural program breaks.

The FCC requires all systems with more than 3500 subscribers to originate a "significant amount" of local programming. Although this amount may include

---

[3] FCC, "Cable Television Report and Order," *Federal Register,* Vol. 37, No. 30, February 12, 1972, par. 118, p. 3269. Hereinafter cited as *Report and Order.*

some of the "canned" programs described, the FCC is referring primarily to programs the cable operator produces himself.

Access Channels

The FCC requires that each new cable system in the top 100 television markets furnish three access channels: one each for local government, education, and the public (public access is also referred to herein as community origination). The FCC prohibits advertising on the public access channel for commercial products and services or on behalf of any candidate for public office. Lotteries and obscene or indecent matter are also prohibited. Beyond these restrictions, the cable operator cannot exercise control over who uses the channel or censor content. The names and addresses of all groups requesting time on a public access channel must be available for public inspection. Operators must keep the records for two years.

The requirements for access channels should not be confused with those for local origination. A cable system may have over 3500 subscribers but not be in the top 100 markets. In that case it must originate programming but does not have to provide access channels. The local franchising authority can demand that it do so, however. While major market urban systems will undoubtedly have over 3500 subscribers, some major market suburbs may not have enough subscribers to require local origination.

The public, the schools, or government bodies may program on other than access channels. But if their programs are classified under "local origination," program responsibility and control shift to the operator. And if leased channels are used, public groups may have to pay rates comparable to those paid by commercial users, unless lower rates can be negotiated as part of the franchise. The access channels, however, are available *free* for an interim period (five years for the government and education channels), after which FCC review will determine future policy.

So far the education and government channels have been little used. Except for a few universities, most school systems still lack the required skills, funds, and equipment, or are simply unaware of the possibilities. The same is true of local governments; one exception is Sunnyvale, California, where new municipal employees are oriented to the community and their jobs via the cable.

The FCC has made provision for expanding leased channels as demand grows; it encourages "cable television use that will lead to constantly expanding channel capacity." The complex FCC policy for expansion—the so-called "n + 1 rule"—is mind-boggling but specific. It states that "whenever all operational channels are in use during 80 percent of the weekdays (Monday-Friday), for 80 percent of the time during any consecutive 3-hour period for 6 weeks running, the system will then have 6 months in which to make a new channel available." *(Report and Order,* par. 126.)

DIFFERENCES BETWEEN BROADCASTING AND CABLE

Broadcasting is an advertiser-supported medium. The networks sell audiences to advertisers on a cost-per-thousand basis, and the programs are important primarily in terms of the audience size they attract. Recently, however, certain audience demographics such as income, education, age, etc., have become perhaps more important because they enable advertisers to target particular groups of consumers. Most advertisers aim particularly at suburban audiences in the 18-to-49 age group, because they consume more than do rural or older audiences.

Broadcasters transmit over wide geographic areas to audiences numbering in the millions, and because only a half-dozen or so channels are available in any one city, air time is scarce and expensive. Since greater audiences mean larger markets for the sponsors and thus larger fees for the stations, each precious moment is programmed to maximize audience size. The system is national in scope (a chosen, not an inherent design), so programs that regularly reach a paltry 5 million viewers will likely be dropped on account of poor ratings—that is, a poor market. It is easy to see why broadcasting is programmed to appeal only to mass taste, with very few exceptions.

Cable can be different. It serves local geographic areas, usually single communities or, in large metropolitan areas, separate "cable districts." Cable programming is supported by monthly subscription fees; revenues from advertising have been negligible thus far.

So many channels are available on cable that it could support a broad range of programs aimed at narrow audiences. As the medium evolves, people are likely to value program diversity at least as much as they value better picture quality. And if you subscribe for the basketball games and I for the weekly experimental film series, it doesn't matter if we watch anything else. Our fees support the entire service, not only our favorite programs.

Broadcasting's most important trait is its highly controlled, well-financed and produced entertainment format. Programs move slickly and smoothly, without a moment lost to indecision or nerves and without so much as a slip of the tongue (even in the news). This "polish" is achieved at the expense of spontaneity and without images drawn from the daily lives of most viewers. Control guarantees predictable quality. Looser approaches may score more memorable moments but will also suffer too many controversial, indecisive, or dull moments to keep a sizable audience. And with air-time so precious, the "uhs" and deadly pauses become too expensive to bear.

So anyone who would produce for cable must ask himself: "Should I, with a fraction of the technical and production resources, without experienced entertainers or writers, try to mimic broadcasting? And although people are so accustomed to it that maybe they expect more of the same, is it worth copying?"

Even if it were possible to match broadcasting quality, the question demands a resounding no. Cable has unique elements that should be reflected in its programs. Programs do not need to be sawed off to uniform lengths. They can be as long or as short as the material calls for. Important issues do not have to be squeezed into 30-second spots and rattled off by agile-tongued newscasters; cable programs can afford to explore issues, drawing upon people less polished in delivery but more powerful as viewing experiences.

EXAMPLES OF PUBLIC ACCESS

As a modest but vivid example of what cable can do, a tape played on a New York public access channel focused on a building superintendent in an area troubled by teenage vandalism. Sitting on an outdoor bench, with kids playing in the background, he told how a group of teenagers had repeatedly smashed his windows. He had three big sons and he originally thought if his sons "woiked 'em ovuh," no more windows would be smashed. But his policeman son-in-law told him to give the boys a chance. They had no place to go and nothing to do. Instead of clobbering them, the super offered his basement as a clubroom. The boys fixed it up with pool tables and magazines, drank beer, and exchanged odd-jobs for rent. Things went well as long as they didn't, as George the Super put it, "allow no girls down there; they didn't mix in my day and they don't mix now." As told by George, the story is warm and funny, personal and alive. It lasts 15 minutes and simply could never appear on broadcast TV in that form. Broadcasters would pluck out the "essential" one or two minutes and thereby ruin the character development and the story.

The point is that cable programming does not require the same kind of control and predictability as network programming. With everyday scenes and realistic everyday people, cable can afford to be boring as well as exciting. Even more than in the early days of live broadcast television, tape-makers can take risks and see what develops.

Differences between cablecasting and broadcasting were further dramatized in a speech by Chairman Pierre Juneau of the Canadian Radio and Television Commission (CRTC, the Canadian FCC). Speaking to the Canadian Television Association Convention in 1972, Juneau told the following anecdote:

> Let me give you an example from a youth concert covered recently by an Ontario cable system. I imagine it drew a large audience, even if only the friends and relatives of the large choir and orchestra watched. They saw their son or daughter or friend performing exactly as it happened . . . in real time. It was well produced—but that's not what was important. Precisely those elements in the program that might have made it unacceptable and boring for a mass audience made it live, authentic and compelling for the involved audience. One moment impressed me: between numbers as the camera lingered, the soloist, a young girl, came on stage with the orchestra and sat down waiting for the piece to begin. There she was, alone, embarrassed, nervous. But we didn't go to a commercial or a commentator to fill in this dead space.
>
> We stayed on the side shot, aware of the clenched hands of the young girl waiting in the centre of the frame. The orchestra tuned up. We still waited with her. Think of what an important and empathetic moment this was in the program for her parents and friends watching. Then reflect on how the professional media would have handled it, and what would have been lost to *that* particular community of interest.

The importance of localism to local viewers should not be underestimated. Look at the treatment of civic issues. Broadcasting portrays such issues in abstractions: unemployment, housing, health are problems far removed from their local context. Issues such as "civil liberties versus civic order," or "freedom versus authority" are not analyzed in terms of frightened neighbors or threatened individuals; instead, people's faces turn into arabic numerals in statistical tables as some high-ranking official discusses "a national problem." This kind of approach breeds frustration.

But cable's eye is on the sparrow; it can plumb local problems and visually depict their impact upon everyday life. Where the broadcasters barely inform, cable can be the instrument for informed action.

Cable can also encourage people to *do* something besides watch TV. Pierre Juneau cited an example in the same speech:

> A Victoria cablecast shows a group of mothers with their children playing constructively with new materials, exploring new teaching games. This amuses the children at home. Like regular TV, the program has a babysitting function. But it goes beyond, for in fact its purpose is to teach mothers [parents] new methods for entertaining toddlers. It's had a good response from adults who have put the ideas to work.

Finally, broadcasting's commercial success prevents it from adequately serving small groups or local needs. By "entertaining" all of us it ignores the ethnic, cultural, and social traits of each of us. Cable can serve locally rooted diversity without bylines, apologies, or illustrations, and without bleaching daily life into the filtered, patterned dullness of the broadcast format.

MAKING ACCESS POSSIBLE WITH VIDEOTAPE RECORDERS

Of course, the basic difference between broadcasting and cablecasting is purely technical. One goes by air, the other by wire. Yet the ramifications of this technical difference are crucial to public access. While broadcast signals are degraded when they are reflected off buildings and airplanes, cable signals do not have these problems. And because they travel directly by wire, cable signals can be weaker to begin with; this means that equipment for producing these signals can be cheaper, smaller, and easier to operate than those required for broadcasting. Recently developed portable videotape recorders (VTRs) are just such a tool.

The most popular battery-operated VTR is the Portapak made by Sony. It is light (about 20 pounds), inexpensive (list is $1650 for a camera/recorder unit), and easy to operate (virtually anyone can learn to use it in about twenty minutes). It uses ½-inch magnetic videotape, which, like audiotape, can be erased and reused. No processing is required between record and playback.

There are basically five widths of videotape on the market: 2-inch tape is the broadcast TV standard, 1-inch is used in closed-circuit and cable systems, and ¾-inch tape is employed by some videocassette machines. Machines using ½-inch and ¼-inch tape are making inroads into the 1-inch market, and are also used for immediate playback situations in everything from golf lessons to group psychotherapy.

Half-inch machines are standardized by the Electronics Industry Association of Japan (EIAJ Type I standard), which means that a tape recorded on one machine will play back on another machine on the same standard—provided both are properly adjusted or synchronized. This sounds simple enough, yet 1-inch machines are not standardized and some models even seem unable to play back tapes produced on another machine of the same model. Exchanging tapes between communities becomes troublesome on the 1-inch format. Of course, if you spend $100,000 for a 2-inch machine or even $5000 for a 1-inch, you get better-quality equipment. Yet with proper maintenance (and support gear) ½-inch machines are usually sufficient for cable production.

The Portapak's ease of operation permits virtually anyone to become a "television producer." In Port Washington, New York, as part of a project run by the public library, 450 people were trained in its use. Tapemakers included teenagers, middle-aged women, young professional men and women, and virtually anyone else interested.

In a San Francisco project "shooting" situation, the "camera crew" consisted of a 79-year-old man and a 13-year-old girl. At Portola Institute, we gave video equipment to hundreds of kids to use in whatever way they saw fit. One group of "academically disadvantaged" kids wanted to do something about billboard construction in their decaying city. They researched the issue, then took our cameras to interview merchants, homeowners, and tenants in their community. The edited tape was taken to a planning commission meeting that was to decide on a permit issue for two giant billboards. All the participants in the tape came to the meeting and saw the commission vote to reexamine the issue.

Videotape's portability and capacity for immediate playback allows the local street to become a shooting set, a TV screen, and a town meeting all at the same time. In a New York City experiment, a video group accompanied by a doctor and nurse went to a busy street and began asking passers-by about venereal disease. A television attached to the Portapak provided live coverage of what was occurring to the gathering crowd. People began to exchange personal experiences about the disease, sharing attitudes on its diagnosis, treatment, prevention, and direct effects on their lives. The doctor and nurse served as technical medical resources.

In addition to exploring the problem of VD and having fun seeing each other on TV, the people involved got a sense of the potential power of a communication system geared to address their needs directly. This tape became part of a comprehensive study of health care in that neighborhood, a study that relied heavily on playback and discussion and on interviews with average people with health problems. When completed, the study would serve as a guide to informed action.

Thus, whether used for cable production or for neighborhood showing, low-cost portable production equipment is vital for making public access a reality, opening up production to new producers, and promising wholly new kinds of programs. Many of these are described in detail later in this report.

II. LIMITATIONS OF THE FCC RULES

At first glance, it might appear that the economics of cable, the technology of video recording, and the FCC requirements are working harmoniously to create vital, effective public access programming. They are not. A deeper look, or a brief taste of experience, will reveal that these elements provide only the bare bones of public access. If interested parties do not press to realize the potential of access programming through hard work and politicking, then other pressures, lack of interest, or ignorance will make public access ineffectual and unimportant.

First of all, we should understand that the FCC rules reflect a regulatory posture of minimum requirements; the burden is on the community to show cause why the requirements should be exceeded. This can be done at the time a newly franchised cable system requests a "Certificate of Compliance" from the FCC. Local groups and citizens can have a voice in this process, as discussed in detail in another report in this series.[4] The FCC will likely bend on many questions provided the supporting arguments are strong.

But as with most cable issues, a better approach is to politick at the franchise authority level *before* the franchise is awarded. While most city councils will tend to provide no more than the minimum FCC requirements in their franchises, community origination can be served better by more "spirited" interpretations, and it is worth keeping in mind that the FCC encourages innovation in the *Report and Order*.

Here are four key problems raised by the FCC rules for new systems in the top 100 markets:

1. Cable operators are required to maintain production facilities within the franchised area for access use, including 5 minutes live time free. However, only minimal equipment—for example, only one studio camera—need be available, thus severely limiting effective programming.
2. With only one channel for public access, regular scheduling of programs or awarding of choice time slots becomes difficult.
3. By requiring systems with more than 3500 subscribers to originate substantial amounts of local programming, the FCC may be inviting conflict between cable origination and public access or community origination.

---

[4] Monroe E. Price and Michael Botein, in Baer et al., *Cable Television: Franchising Considerations*, Crane, Russak & Co., Inc., New York, 1974.

4. Funding: the FCC has left the financial problems of access up to each community and restricted the financial burden a community can impose on a cable operator through franchise fees and the like. The FCC also prohibits mandatory subsidy of access by the operator.

These four points are discussed in turn under the subheadings below.

TIME AND EQUIPMENT LIMITATIONS

The FCC requires "one dedicated, noncommercial public access channel available without charge at all times on a first-come, first-served nondiscriminatory basis . . . but production cost (aside from live studio presentations not exceeding 5 minutes in length) may be charged to users." *(Report and Order,* pars. 121, 122.)

The problem here is best illustrated by an actual program experience. A public access or "open" tape from Berkeley, California, showed a nervous young man facing a single camera in a bare studio. The words "Open Channel, Open Channel" framed his head in huge letters, like a T-square. He talked about the bombing in Vietnam. In some sense, this unfiltered opinion on a crucial issue made historic viewing; but the format undermined, if not obliterated, the message. If this were typical, the influence of public access would be very slight.

The tape at best afforded catharsis for its speaker. Would-be viewers cannot be expected to contend with so narrow a presentation format. The "Open Channel, Open Channel" frame around the speaker seemed to be an apology either for what was said or for the fact that such a channel is required in the first place. The stolid camera set-up and the nervous, didactic speaker almost invited the viewer to dismiss the message. The point here is not that access programs must mimic broadcasting, but that even a 5-minute segment should be regarded as a program. It must reflect the producer's equal concern for the way a program comes across as well as for what is said.

The fault lies as much with imagination as with economics or regulations. But the importance of program technique in getting a message across is a sophisticated concept, and many poor community groups are likely to regard the free single camera as sufficient—particularly since costs for more elaborate studio time often exceed $50 an hour and cheaper portable equipment is not always available. (In a pamphlet published in August 1972, the National Cable Television Association recommended that operators make studio time available at cost to access users.)

The problem is that the term "access" means only the ability to get on the system; it says nothing about what to do when you get there. There are two reasons to get on the cable. One is pure catharsis—to release the frustration of silence; the other is more demanding—to present a message for all or part of the community at large, in hopes of being heard. But the capacity to be heard on television involves more than the ability to speak. It requires at least a minimal understanding of the language of television.

SCHEDULING LIMITATIONS AND VIEWING HABITS

Access programs, like those on broadcast TV, are best served by regular scheduling. Broadcast programs appear with regular or "star" performers in regularly scheduled spots, or as specials at times scheduled regularly for specials. Audiences build up viewing habits based on these patterns. Vary a show's day and time from week to week and its audience share will plummet. Time-slot programming serves economic rather than aesthetic purposes. Timing on cable need not be as rigid as for broadcasting,[5] but regular scheduling is very much a part of building an audience and must be taken into account.

Regularity is particularly important for specific information programs (e.g., health and consumer information) and for specialized audiences. In New York City, for example, the Deafness Research and Training Center provided news and programs for hearing-impaired people in prime time, all in sign language. Without a dependable regular time slot, the audience (which does not regularly watch TV) would have missed much of the vitally needed information and entertainment.

Programs can be scheduled regularly on the basis of subject (e.g., 7:30 Tuesday is employment opportunities hour), or by program producer (e.g., a local video production group is featured Wednesdays at 8:00). In addition, one-time programs or "specials" might request a particular time slot. Scheduling of this sort will be difficult, and it may be impossible to satisfy the demand, if there is only one public access channel. Possible solutions may be found among the list of alternatives at the end of this subsection.

Viewers often tune in before their favorite programs come on, and leave their sets on afterward, thereby seeing a good many programs they would have missed otherwise. As applied to public access, this habit would warrant a separation between programs that hope to build up regular audiences and random "one-shot" programs. This separation has nothing to do with program content, nor should it depend entirely on production resources (see "Funding Limitations" below). It merely reflects commitment to regular programming.

Program separation calls for two assignments of channel time. One is for the 5-minute segment prescribed by the FCC (such as the Berkeley tape discussed earlier); the other would award regular time slots for extended periods on a first-come, first-served basis. Specials would take either an available time slot or trade spots with a regularly scheduled program.

Another important viewer habit concerns the channel itself. Many viewers become attached to particular channels and ignore the rest. This may put access programs, aired on a "special" channel, at a competitive disadvantage. One solution would be to intersperse access programs with popular local origination, such as local sports or movies, to take advantage of audience carry-over.

The cable operator currently handles scheduling on a first-come, first-served basis, for either regular or random programs (e.g., the Berkeley tape). (Other methods, such as the creation of local public access boards, will be explored later.) Even on this basis, there are several methods for allocating channel time to permit regular scheduling. Some methods emphasize regular programs, perhaps to the detriment of less organized productions. This matter of choice will recur throughout the

---

[5] BBC programs are only as long as their producers feel they should be. They often begin at odd times like 9:17 or 11:43.

report in discussions of equipment, program style, organization, and funding. Should access emphasize simply getting people on the air, or should some flexibility be sacrificed to accommodate the audience? It is possible to do both, but at the cost of more effort and more funds. It may be necessary to restrict completely open individual access in favor of organized community groups and regular programs. The choice is yours to make and some scheduling alternatives are listed below. *Whichever method the community decides to use, it should be stated in the franchise, and the community should be prepared to defend it during the FCC certification process.*

1. Regular programs or carefully prepared tapes can be interspersed with short segments throughout the day. Doing so may reduce the effectiveness of back-to-back regular programs, but may acquire audiences for the random programs.

2. The program day can be divided between random programs in the morning or afternoon, and regular programs in the evening. This will depend on the supply of each type.

3. Channel time for regular or random programming can be borrowed from unused periods on education or government channels. This alternative, which requires education or government sponsorship, has not yet been attempted and in fact is useful only for particular programs.

4. A second public access channel for regular programming can be written into the local franchise ordinance, although the community may have to defend it in the FCC certification process. The New York City franchise (written prior to the FCC *Report and Order)* requires two public access channels to "encourage differing uses.... On one Public Channel ... there shall be an opportunity to reserve a particular time period each week, or several time periods each week, in order to permit the user to build an audience on a regular basis.... On the other Public Channel ... there shall be no multiple time reservations, in order to permit a user with a single program and users with relatively last-minute requirements access to prime time periods."

To prevent a few major nonprofit organizations from dominating reserved channel time, the New York City regulations impose limits of 7 hours a week, including 2 in prime time, per person or organization. Of course, these limits can be adjusted according to the local demand for access time. (Studio rates can also be moved up and down to encourage or limit extended use. Rates for the third hour or the third time of use, for example, might be lower than for the previous two—or higher.)

POTENTIAL CONFLICTS BETWEEN CABLE-OPERATOR AND COMMUNITY ORIGINATION

The cable operator is the wrong man to saddle with responsibility for access. In a city where the FCC requires both local origination (cable system has more than 3500 subscribers) and public access (city is in the top 100 markets), an operator's ability to support access is limited. His dual responsibility can cause conflict over the use of studio facilities and personnel and interfere with revenue-producing programs and services.

An operator can rent his studio to commercial customers, sell advertising for his programs, and try to attract subscribers with his own origination. His efforts have not yielded much money so far, but he believes they will in the future and he programs accordingly.

It has been said over and over that because cable gets regular revenues from its subscribers, it does not need to worry about building large audiences for its programs so as to attract advertisers. But subscriber revenues often go toward meeting debts incurred in building the system, and only skeleton budgets are possible for origination efforts. Program directors often must look to advertising to help meet program expenses. Ad revenues thus far are negligible, but the point is not whether in fact substantial revenue is brought in, but rather is the cable operator's interest in mass-audience or "safe" programs, such as local sports, that bring in as much revenue as possible for a given program slot.

U.S. cable operators, then, despite competition from broadcast channels and limited production funds, still imitate broadcasting as best they can. They are less formal and polished, more flexible, and perhaps more concerned with local needs— but they are still producers in pursuit of "quality" programming to attract large audiences and advertisers.[6] In the top 100 markets, their role as producers puts U.S. cable operators in a bind. As the FCC envisions public access and community origination, the cable operator provides the equipment, advice, and assistance for access productions. Ideally, he meets all requests for equipment and assistance, perhaps generating new requests with enthusiastic promotion of access. But in actuality, an operator can jeopardize a potentially effective access system through lack of resources or interest, or conflict with his own productions.

Required by the FCC to maintain a studio for access use, an operator with more than 3500 subscribers must also use his studio to meet the local origination requirements. Public access programs bring him no revenues and cost him valuable time he could spend on revenue-producing origination. And possibly, but not necessarily, his "safe" programs may attract more subscribers than access will. Thus, where his dual role causes conflict, access is likely to be the loser. Few operators will do more than contact local civic groups or announce that access time is available—hardly the depth of promotion needed to spur local interest. Perhaps for these reasons, the NCTA recommends that system operators seek community advice and assistance to help promote and develop access. Therefore community groups should encourage cable operators to involve community people in administration of the access channel.

The potential operator versus community origination conflict varies from town to town according to community demand for access. In isolated or smaller, more homogeneous communities the demand for access is often very low. While local identity is strong and people want to see locally oriented programs, they may not feel the need to do the programming themselves. Production monies and volunteer time are limited, and cultural diversity is not as pressing a demand. Local programming responsibility is often entrusted to the cable operator in such places.

In large cities, the sense of community often derives not from geography but from common interests and ethnic and racial bonds. The cable operator (or any

[6] In Canada, operators are happy to have community groups meet their programming needs for them. In fact, community origination supplies about as many hours of viewing as operator origination does, according to a survey of *The State of Local Programming*, Broadcast Programme Branch, CRTC, July 1972.

"outsider") is rarely viewed as the right person to program for these diverse interests. People want to represent themselves in programming, therefore placing a substantial burden on the cable operator for equipment, assistance, channel time, and perhaps even operating expenses. Many urban cable systems (such as the one in Reading, Pa.) enthusiastically meet these requests, if only to build good will in the community and acquire a showcase they can cite in their other franchise applications, or as an inexpensive means of filling their local origination channel. In New York City, Sterling Manhattan (owned by Time-Life) is building separate access studio facilities. But not all systems are so well endowed and many will barely meet minimum FCC specifications.

The problem becomes most acute in suburbs where both origination and access are required. Suburban communities are often just large and diverse enough, with enough interested citizens, to generate strong demands for access. At the same time, the potential number of subscribers in suburban markets is often too small to provide the incentives a cable operator needs to serve both roles adequately.

Of course, the operator's own cablecasting programs can also be an important local voice, and bring greater resources and perhaps more polished skill to local programming. But unless the conflicts with community origination are solved by access programmers somehow getting hold of extra facilities, additional personnel, or funds, operator origination will succeed at the expense of public access. In fact, limitations of funding and volunteer talent in many communities may prevent either operator or community origination from succeeding. In other communities, able to sustain one or the other but not both, competition for money, talent, and time could cripple both.

Finally, operator and community origination differ along more than mere economic or commercial lines. As their intent is different, so is their capacity to involve people in the production process. With the advent of portable video gear, community origination can be done anywhere, by anyone. The neighborhood can be the studio, and the neighbors producers and players. Access producers with portable equipment can reach situations and people a cable operator could not or would not cover because of lack of time, inclination, or support. Or he may be unable to probe deeply into local life simply because people are wary of him. A man may talk intimately to friends holding cameras, but clam up before the cable operator's cameras.

Also, most operators confine the bulk of programming to the studio, promising network-style control and predictability. On location, the large and imposing studio cameras do little to improve mobility or flexibility.

FUNDING LIMITATIONS

Funding is the most serious obstacle to access. If neither the franchising authority nor the cable operator supplies funds, access must resort to the hodgepodge of money-raising schemes so common to other public interest projects. With advertising prohibited and subscriber revenues signed over to the cable operator, access has no regular means of support. Most of the projects described in this chapter are foundation supported, but that source may dwindle in importance as public access expands to hundreds or thousands of communities. Foundation funds cannot support access programming indefinitely on a nationwide scale.

Access costs have three major components: equipment and tape—perhaps the most difficult to come by; personnel; and operating expenses and promotion. Each resource can be acquired separately—equipment from here, an office from there—or funded by a single source. While individual groups can always borrow production gear from a church, school, or government agency on an ad hoc basis, community-wide, day-to-day availability is limited. If equipment cannot be rented (or "borrowed" permanently) or purchased out of access funds, only well-endowed groups may be able to produce tapes. Equitable access demands the provision of facilities for community-wide use.

The most coveted source of access funding is the cable system and its revenues. Both cities and community organizations have their eyes on that pot. To guard against excessive taxing of cable revenues which might burden the operator and retard cable development, the FCC generally prohibits a direct tax on the system to support public access. The cable owner pays a franchise fee to the municipal authority of not more than 3 percent of gross subscriber revenues. To increase the franchise fee, or to require direct subsidy of access by the operator, demands a special showing before the FCC. The municipality *and the operator* must demonstrate that such increases are necessary and do not hamper the system. (See the discussion of special showings in Sec. III.)

Still, attempts are under way in many communities to directly attach some portion of subscriber revenues for access. The amount is usually expressed as a percentage of revenues (perhaps 1 or 2 percent), a per-subscriber fee, a flat annual sum, and so on. In other communities the operator is being asked to provide equipment, operating materials and expenses, and personnel. In Berkeley, for example, community groups are asking for $40,000 worth of equipment and operating expenses for the first year of access, to be followed by $15,000 each succeeding year. Since all these demands require FCC certification, it takes political skill to secure the operator's (and the municipality's) cooperation.

The municipal franchise fee, earmarked principally for the local regulatory program, is another target. The FCC's only concern is that the fee does not exceed 3 percent. If it does not, a part of the franchise fee can be spent on public access without FCC approval. The problem is to decide whether access deserves priority over the local regulatory program or other community needs (such as day-care centers) and, if it does, to determine if the proponents of access have enough political clout to shake funds loose from a revenue-starved city.

For access, the best recommendation is to start with the cable operator and the municipal authority, and go on to other fund-raising efforts from there. One approach is to obtain equipment and tape from the cable operator (for tax purposes, he often would rather supply fixed assets than cash), and personnel costs and operating expenses from the city. Although this plan has yet to be implemented, it is the core of the public access approach in Santa Cruz, California.

Indirect and direct advertising can be another source of funds. Although the FCC prohibits commercials on the access channel, businesses can contribute to an access fund. Tax-deductible gifts can be solicited for a nonprofit access corporation, and the donors mentioned in a monthly access-program guide either as a group or as the sponsors of particular programs. And advertising space can be sold for the guide. These and other indirect advertising schemes, such as a public access page in the local newspaper, paid for by business but advertising access, can be tried in combination. The problem is that access may be too controversial for business to support, or that donors may try to limit support to particular programs of their choice. This too can be played to advantage, provided other support is found for programs business is unwilling to fund.

Sponsorship of a specific program cannot be mentioned on the air, but again the monthly program guide or the automated program announcement channel can provide credits along with schedule information. These and other methods can be explored to take advantage of, for example, a drug company's willingness to fund drug abuse programs. It must be made clear to any contributor that while his gift may bring tax or publicity benefits, it does not give him any measure of control. For this reason, the best approach, if it can be worked out, is to pool all gifts in a general fund.

Federal and state programs are a likely source of specific program monies for health care, law enforcement, welfare information, and the like, but the government is unlikely to pick up the tab for basic equipment and personnel for more than a few experimental systems.

Cake sales, marathon walks, United Fund or other attempts at indirect support from local citizens are also possible.

Finally, if no other source of revenue is available, try to sell ads. This may defray production costs for a program carried on the local origination channel or it may enable you to lease a channel. This is not exactly public access, but without funding it may be the only access possible.

III. ORGANIZING FOR PUBLIC ACCESS

BUILDING SUPPORT

The first step toward almost any form of public access is to inform the community about cable and arouse public support, particularly from groups actively involved in community affairs with concrete needs for access services. Promotion is vital for getting people to take an active role in the franchising process, encouraging use of production facilities, or building up an audience for access programming. Radio, television, newspapers, bulletin boards—all media can be used. In addition, many access groups have developed innovative organizing approaches more specific to the cable question.

The Watts Communications Bureau in Los Angeles, one of the first community organizations interested in cable, has been working for three years to gain community support for its bid to own the franchise in a predominately black area. A division of Mafundi Institute, a community cultural center, the Bureau has been stressing the programming and revenue advantages of community ownership.

WCB is training young people in video and sending them out to cover local events and issues. Bureau staffers can show the resulting programs to local organizations or perhaps to an influential minister. The message is that he too can use this service for his communication needs. The sight of local people operating video cameras also prompts questions from citizens, providing a perfect opportunity to explain cable and community access.

The Watts Communications Bureau also works with the Watts Writers Workshop, local dance, music, and theater groups and other local people in the arts. In addition to developing a supply of talented producers, the Bureau is stockpiling tapes to prove to people that community-owned cable will present black artists in black productions. It also demonstrates that programming will not be exclusively issue-oriented but will offer a kind of entertainment, particularly for black audiences, not now available on television.

To develop support for its franchise drive, Mafundi has cosponsored two cable symposia featuring national cable figures and showing locally produced tapes. The sessions went largely unnoticed on the grass roots level (as expected), but influential city councilmen, local artists and community organizers, popular radio personalities, and prominent black businessmen attended and were excited by the prospects.

Mafundi's latest tack in spreading the word about cable is a survey of community television habits and interest in cable. The survey provides the opportunity to explain cable and asks, among other questions, whether people would pay $5 per month for increased television service and whether they would watch local programming. The first question is designed to demonstrate that in black ghettos, too, cable will sell. The second question—revealing that a high percentage of viewers would be interested in local news, math and reading programs, and so forth—will buttress future arguments for strong commitment to local programming from whoever owns the system.

In Santa Cruz, California, Allen Frederiksen concentrated his campaign for a public access corporation on the youth of Santa Cruz (see "Public Access Board" in Sec. IV). He recruited high school students by doing frequent videotape workshops for media classes. Then, armed with posters drawn by his high school media students, Frederiksen would set up a VTR in the student union of a nearby college. Janis Joplin and Mick Jagger tapes (recorded off the air) blared from monitors, attracting passers-by to take a closer look. Frederiksen casually interspersed a prerecorded tape about public access in Santa Cruz between the rock star tapes. He passed out leaflets, answered questions, and showed examples of access tapes, then moved on to the next site—another school, the town square, or perhaps a local youth hangout.

These efforts, coupled with a talent for newspaper and radio publicity, gave Frederiksen command of a vigorous army of several hundred young people to gather petition signatures, produce tapes, and threaten uncooperative councilmen with the wrath of their newly won power at the ballot box. His techniques may be useful in university communities around the country.

The Community Video Center in Washington, D.C., sends out literature explaining video and cable to community organizations, offering to train members and lend equipment. Video presentations follow mailings where possible. Washington is not yet cabled, but growing segments of the community are realizing their stake in the issue. Overcoming the initial barrier of the technology goes a long way toward developing an understanding of cable access.

The Washington approach of working through community organizations (anything from the PTA to the local women's health collective) not only offers the easiest way to contact people directly but seems to strike at a basic access need as well. Organizations must communicate with current members and with the community at large to attract new members. They must spread their message, and cable promises to be more effective for their aims than mailings or occasional radio announcements. Organizations may have the budget, size, and discipline to aid in any public campaigns. They are also likely to be the first and most consistent access users. And their involvement in community issues indicates that they will not take an interest in cable for its own sake, but will work to make it an effective instrument for grappling with local problems. This is vitally important. Access is useful only to people or groups with a need to communicate. These groups, although active in community affairs, may not take to access as readily as videophiles, but they are central to access development as a channel for change. So organizers are encouraged to recruit their support and help them shape the system to their needs. For once the initial excitement dies down, these groups will carry the access load.

These are only a few examples of organizing for access. The important thing to remember is that most people in your community may know little about cable and its potential and even less about doing their own programming. Effective public access requires informed public support—the earlier in the franchising process, the better.

THE FRANCHISING PROCESS

The franchising process is the crucial period for citizen involvement in cable. This is when municipalities determine what sort of system they want, solicit proposals from interested cable operators, and award the franchise. Citizen groups must first of all lobby for a franchise agreement responsive to local needs. This means presenting your recommendations to municipal authorities and backing up your recommendations with data and the political power of public support.

In Berkeley, organizers well informed about cable sought to build a strong case for public access (elsewhere the prime issue is often community versus private ownership). They approached a host of community organizations, asking if public access would be helpful in their work. In many cases the organizers had to begin at the beginning by explaining cable and its potential, of which some groups had scarcely even heard. Other groups, notably the Third World Media Coalition, knew all too well the potential utility of access to television—because they had been denied it until now. In their presentation, the organizers pointed out that not only is public access vital to a great number of citizen groups, but that equipment, funds, personnel, and flexible channel scheduling are needed to meet the demand. The situation in Berkeley is currently under study. The major question is whether the proposed changes in the franchise ordinance constitute a breach of contract with the owner of the already-operating cable system.

Most campaigns, supported in this manner and with the political leverage to back them up, should be successful. Although a well-documented case can be helpful, it is not likely to be enough.[7] Political power as represented by widespread public interest may be much more important in dealing with municipalities at the franchising stage. Cities may be tempted to make concessions to private interests because they are so perennially preoccupied with finding short-run sources of municipal revenue. Sometimes only an aroused citizenry thumping noisily on ballot boxes can distract their attention. And in the case of cable, it may do some good to point out to municipal officials that cable systems are unlikely to be highly profitable in urban areas for some years to come.[8] Franchise fees will not be a gold mine for a while.

Another approach to franchise politics is to bargain with a prospective cable operator. You can exchange your political leverage with the franchising authority for perhaps a percentage of ownership or at least production facilities and operating expenses. In the past, cable companies have taken on influential local lawyers and businessmen as partners in the local system, surrendering equity for their help with the franchise. The same can be done for a public access board.

In short, the franchise process is the time to lobby, because it becomes a sticky legal matter to amend an agreement once it is signed (see the passage on "grandfathered systems" below).

---

[7] In San Mateo County, California, a well-documented case by a black community group was ignored; they simply were not a powerful enough political force by themselves.

[8] See the discussion of "System Profitability" in Baer, op. cit., Chapter 3.

SPECIAL SHOWINGS BEFORE THE FCC

Even if you push your demands for flexible channel time, free studios, and the like, past the franchising authority and the cable operator, they still must be certified by the FCC—as part of the "certificate of compliance" that the cable operator must receive from the FCC before he can begin operations.[9] Because these provisions may exceed FCC specifications for access, a petition or special showing may be required before the Commission. However, the Commission emphasizes that "we are entering into an experimental or developmental period. *Thus where the cable operator and franchising authority* [both] wish to experiment . . . we will . . . consider the appropriateness of authorizing such experiments, to gain further insight and to guide future courses of action." *(Report and Order,* par. 132; italics mine.)

The Commission, then, is willing to approve local solutions to access issues (e.g., using franchise fees to support programming of the access channel), but not at the expense of the cable operator. The fear is that franchise awards will be based on bidding-war promises of extra channels or lavish facilities.

Thus to gain Commission approval, or to even persuade the franchise authority or the cable operator to petition, you must make a strong case. Decisions on the proper number of channels, scheduling needs, and other access questions (equipment, operating expenses) demand a sense of community requirements. The FCC must be convinced that the funds and facilities are needed and will be used, and that the financial burden placed on the cable operator is not unreasonable.

While survey data, sample tapes, lists of prospective users, and statements of need may answer the first FCC question regarding demand, the second question is more complex. The cable operator may agree to your terms to get the franchise, only to complain (covertly) to the FCC that he was unduly pressured. There are basically two ways to discourage this practice. First, it should be explicitly stated that if the FCC disapproves the agreed-upon terms, the municipality may reopen bidding for the franchise. Thus an operator will not agree to more than he feels can be successfully supported at the FCC certification process.

A second approach (as yet untested) is for a municipality to draft a detailed franchise in advance of bidding competition, so that all applicants would have to meet the same fixed conditions. The winner would agree in advance to support these provisions before the FCC. This procedure would allay the FCC's stated fear of bidding wars and illusory promises.

These are the demands of the certification process: to support your stated needs and to demonstrate that they do not constitute an unfair burden on the operator. Again, both the operator and the franchising authority must petition the Commission and letters of community support will be helpful.

ACCESS ON GRANDFATHERED AND OTHER SYSTEMS

The FCC rules for access currently apply only to new systems in the 100 largest

---

[9] To carry broadcast signals, which are regulated by the FCC, cable systems must comply with all Commission rules. For a complete discussion of the legal issues involved, see various other reports in this series, but especially Price and Botein, in Baer, *Cable Television;* and Steven R. Rivkin, *Cable Television: A Guide to Federal Regulations,* Crane, Russak & Co., Inc., New York, 1974.

television markets. For systems outside the top 100 markets that began operating after March 31, 1972 (the effective date of the rules), municipalities can require the same minimum access provisions. Systems operating prior to the effective date of the rules have five years in which to meet the new requirements. These systems are said to be "grandfathered"; but if they take advantage of the new rules so as to carry additional broadcast channels, they must then provide at least a public access channel.[10]

Grandfathered systems present a tangled web of legal and political problems to anyone interested in updating cable services. The older franchise agreements pre-date community awareness of cable's potential and, of course, current regulatory and technical developments. Thus most systems were built with only a 12-channel capacity, and community origination was not provided for. The operator filled his channels with broadcast signals and his own origination, and access was left without channel space. In 1972 about 400 U.S. cable systems had a capacity of 5 or fewer channels. Such systems generally had no access capacity for either local origination or public access.

In Santa Cruz, Allen Frederiksen's student army ran into charges leveled by the cable operator, and supported by the city attorney, that implementation of their demands would constitute breach of contract. Frederiksen has been forced to organize a petition drive to put his demands for public access on the ballot. Even if affirmed by the voters, the case may still be taken to court to settle the contract dispute. Given these problems with grandfathered systems, the need for a proper franchise agreement at the outset becomes obvious.

Despite these examples, grandfathered systems have their advantages. With several years of operation behind them, many systems have paid off original construction debts and are beginning to show a profit. While these systems are better able to support public access, their success lessens the need for the good will and public-relations value such support might give them. In some cases, however, support for access provides glowing credentials for franchise applications in other cities, and as such is an economical public-relations scheme. Grandfathered systems are not required to petition for FCC approval of practices that go beyond the rules, and some of these practices are beneficial to the public. In New York City, for example, there are two access channels, and the operator in Reading, Pennsylvania, provides funds and equipment for access even though his franchise does not require it.

In short, the willingness of grandfathered systems to offer the services required of new systems depends on the limits of the franchise, and more important, the good will of the cable operator. Community groups are well advised to negotiate with the operator before resorting to legal action, subscriber boycott, or other such tactics.

---

[10] For more detailed discussions, see Baer, op. cit., Chapter 8, and the legal analysis in Rivkin, op. cit. The FCC regulations concerning grandfathering appear on pp. 3267-3268 of the *Cable Television Report and Order,* which Rivkin reproduces in full.

IV. MODELS FOR ACCESS

Securing a favorable franchise is only one step in the development of effective public access. The next step is to create a solid framework around which access can be built. The sparse FCC rules describe an access cast with only two characters—an administrator and a user. The administrator supervises channel operations and each individual user worries only about his own message. The rules are silent on issues relating to the operation and effectiveness of the access system as a whole. Thus, issues whose solution would improve the effectiveness of the entire system are never mentioned—for example, getting people to use the channels, promoting the programs, raising funds for production, and training people to use the equipment. The result is that only well-funded, organized groups can readily take advantage of access.

But there is a bright side to the picture. The absence of rules and guidelines also means freedom to experiment and adapt access to local needs. Community groups are trying out various approaches inspired by a variety of different ways of looking at access. Some groups are exploring new methods of channel administration, such as establishing a nonprofit corporation to administer the channel or affiliating with a local institution that would not only run the channel but also provide programs and training. These groups are primarily interested in shaping the total effect of the access channel. Others have more specialized interests. Some projects emphasize getting as many people as possible on the air, while others focus on the issues to be explored. Some facilitator groups help access users produce the best-quality tapes possible (doing it for them if necessary), while others concentrate on involving people in the "process" of video.

This section presents several ways to establish and operate access; you must select among them according to your goals and resources.

FACILITATOR GROUPS: NEW YORK CITY

Facilitator groups act as independent advisors to users and to the system as a whole. They cooperate with the cable system, but they do not establish rules or assume administrative functions, and thus do not require FCC certification. They can be nonprofit or commercial, full-time access advocates or part-time production groups. Some have their own goals; others simply help users pursue theirs. Most communities will have room for a variety of facilitator groups, such as the three kinds described below that operate in New York City.

Open Channel

Open Channel is a nonprofit corporation supported primarily by grants from arts councils in New York City and New York state and from a number of small foundations. Organized to develop and extend public access, Open Channel deals with city and state regulatory bodies, negotiates with cable companies, and raises funds for access productions. It owns very little equipment, preferring to spend its money on personnel and the overhead expenses resulting from its campaign for public access. Out of a first-year operating budget of $150,000, less than 10 percent went for equipment.

Open Channel is almost exclusively a promoter, making potential users aware of cable and access in particular. It informs people about available channel time and production facilities and preaches the benefits of access programs.

But promoting is not organizing. Promotion seeks to popularize a general concept through indirect contact or publicity. It is geared to attract mass attention and passively relies on the listeners to respond without being approached directly. Organizing involves more direct, active persuasion. It bends to the individual needs of each user. Consequently, organizing is smaller, if more intense, in scope. Open Channel does some of both. Its staff has neither the numbers nor the experience to permit a grass roots organizing approach. Instead, the limited organizing effort is focused on groups that can have a "rippling effect" in the community. Neighborhood newspapers, influential churches, powerful neighborhood organizations are encouraged to make programs regularly so that an important local base for access experience and perhaps facilities will be established. This approach is part of Open Channel's overall policy of minimizing the random remarks of individuals and concentrating on groups involved in community problems and culture.

As a promoter, Open Channel uses every popular vehicle available to spread its work. According to its founder and director, Thea Sklover, "We use full-scale advertising, hand-outs on street corners, articles in newspapers, appearances on radio and television, talk shows, anything that will get the message out." Future plans call for a saturation advertising campaign in supermarkets, laundromats—any place frequented by large numbers of people.

Open Channel is also pushing for neighborhood studios, hoping that readily available equipment and a limited neighborhood audience will make it easier to attract users and to show programs on strictly local information and culture—for example, Chinese language programming produced at the Chinatown facility.

Another Open Channel goal is the creation of video viewing centers at schools and churches for people not yet on the cable or too poor to subscribe. They hope to generate neighborhood awareness by expanding this concept to include bars, laundromats, and other natural gathering places where local material might be received.

Open Channel's efforts are designed to excite immediate interest in public access. Its programs follow a similar tack. Technically polished and well produced, they reflect strong interest in the message and its audience. Open Channel believes that "soap box" access—simply getting people on the air—sells public access short. You need not be slick, its staff members concede, but amateurish, shaky camera productions will turn off audiences attuned to polished broadcasting. High technical quality will engage the audience's attention and fresh, local content will keep them tuned to the access channel. To ensure success, Ms. Sklover has managed to attract volunteers from the large pool of professional talent available in New York City to help groups produce their programs. But the producers subordinate their aesthetic concerns to the group's message. The group supervises editing, and they can preview the final tape before it is aired. As soon as they are made, tapes are usually replayed immediately for the participants. Additional showings are arranged for friends and neighbors to get their reactions and to involve still more people in the process.

But production is not the main function at Open Channel. (Even New York City has just so many producers with time to spare.) They produce because a supply of technically good programs helps attract audiences and user groups to public access, and demonstrates its potential to other cable communities. As users develop their own production skills, Open Channel plans to move from production to training.

Although always part of their publicity, training never quite fits their production methods. Open Channel's concern is for good programs. Their productions, particularly one-time events or special situations, do not allow for adequate training. The choice is essentially between product and process; that is, do you want "high quality" programs regardless of how they are produced? Or do you want to teach people to make their own tapes, worrying less about how the final program may look? Thus far Open Channel has opted for "high quality" tapes—perhaps one reason its volunteer talent pool continues to grow.

Recently, however, small-scale training programs at a few schools have mushroomed into contracts with the New York City Board of Education and other education money sources.

The schools are well endowed with equipment (often used poorly if at all) and the training programs hope to foster use of video within school walls as well as for cable. In addition, training contracts may provide Open Channel with an on-going source of revenue, should foundation support suddenly dry up.

Finally, to keep things in perspective, Thea Sklover points out that training may be a misleading issue. Many groups will have neither the time nor the inclination to produce their own tapes. Others will want the kind of production quality that only professionals can offer. These people will need a production service to translate their ideas to videotape. Open Channel's goal, then, is to make sure these and other access services exist, and that everyone knows where to find them.

Alternate Media Center

"I went up to Washington Heights and was talking with a guy and said, 'What bothers you? What are your problems?' and he said 'Garbage,' and my reaction was that this guy was evading my question.... It wasn't until I'd ... really got into the neighborhood that I knew it was a real problem. I'd wanted him to talk about some deep incredible problem about life on a gut level and he wanted to talk about something that really bothered him and that was the garbage."

— JOEL GOLD, Alternate Media Center

The Alternate Media Center, housed at New York University, offers a completely different approach to public access. The Center is not interested in promotion, although its directors, George Stoney and Redd Burns, often find time to speak on conference panels and lobby for funds at foundations and large cable companies. It prefers the slow, unpublicized, long-range process of involving people in their communities. The Center's goals go beyond public access television. It uses television, or more accurately, video, to enable people to take charge of the information that directs their lives, and to open up neighborhoods and towns to vibrant, new patterns of communication.

To achieve this goal, Alternate Media functions as a resource center. It provides equipment, skilled technical personnel, advice and assistance, and organizing experience. Equipment and assistance are available both to people the Center seeks out and those who wander in unannounced. In tapes like "George the Super" (described earlier), students or staff from Alternate Media wield the camera. In other situations, the cameraman may be anyone involved in the project. Training and production become part of the same experience. Because the goals are long range, the Center attaches more importance to the process of using the equipment than to the quality of the finished tape.

Although the Center itself has produced several fine tapes, in a typical project it casually introduces the equipment to a group unfamiliar with video. Group members get "hands on" experience as they use the equipment to explore their surroundings, taping family and friends, shopkeepers, and so on. "Play" is encouraged. Eventually, the group begins to point the camera inward, focusing on itself, its purpose, and the issues it is concerned with. Cable finally becomes involved when the group decides to seek a larger audience. Only at this moment do questions of video technique and program style enter the picture. Prior to this, the Alternate Media resource person simply gets to know the group and its central concerns and serve as a minimal technical aide, using the equipment when necessary, offering advice, but at no time imposing a producer's contraints on the process. Once the tapes are made, the Center encourages the feedback process as a vital part of the project. Too often, conventional TV merely fills up channels with one-shot statements of problems, leaving the audience frustrated and only superficially informed. They are not motivated to act on the problem, partly because no opportunity for action has been presented. The audience can do little more than roar at the tube or continue watching in passive silence. By conducting public viewing sessions and continuing to tape on a specific, local issue, Alternate Media tries to bring about not only a greater understanding of the issue but also a sense of participation in its solution.

For example, Alternate Media was present at a community planning session on schools in Greenwich Village. All 24 hours of the sessions were videotaped and cablecast with only about an hour's delay. Participants were able to catch up on missed sessions either via the cable or by reviewing the tapes on playback decks provided by the Center. In addition, TV monitors carried the sessions live out on to the sidewalks, drawing passers-by into the discussion.

Perhaps the tapes most often produced at the Center are straightforward interviews with average people about their concerns. These tapes hope to present everyday images with which a wide range of people can identify. By contrast, Arthur Bremer, convicted assailant of George Wallace, wrote in his diary about the traumatic realization that his family was not like the one portrayed on "The Donna Reed

Show." Alternate Media believes television should reflect the diversity of life, rather than idealize one narrow segment.

Taping at this level of intimacy requires trust between subject and production crew. The subject must believe that the tape—his image—will not be synthesized into a two-minute spot on the evening news. People of all persuasions are very sensitive to being used by the media as stereotypes—e.g., "the typical hard hat" or "long-hair." They must be assured that the tapes will be used only with their permission and that (out of respect for them as people) edited material will not be cablecast to an audience of their friends and neighbors without their permission. Although it may seem cumbersome, the latter stipulation poses problems only to those plagued by the demands of a broadcast schedule. Public access programmers should take the extra time to consult their subjects.

Alternate Media believes its participant-oriented television offers people not merely information or entertainment but an opportunity for involvement. Just as merchandisers encourage you to consume, access channels can encourage you to participate in a public forum, a cultural event, or the county fair.

For example, Joel Gold of Alternate Media taped a demonstration held to point out the need for a traffic light on a busy New York street. The tape played on the public access channel, and as a result, more people from the neighborhood attended a meeting about the light. Joel taped that meeting in turn for the access channel, and the process continued until the local people, without an organizer, formed an action group on their own. Video reinforced and strengthened the group's position. Budding neighborhood leadership was brought out and permitted to learn by experience. The traffic-light group has since gone on to other issues.

This example points up the advantages of neighborhood studios and locally oriented programs. It also demonstrates television's potential for dealing with local issues (albeit in this case a small one). In its various projects around the country, Alternate Media has dealt with more dramatic problems, including racial antagonisms; and the material, although objectionable to some, is never watered down or qualified to make it palatable to all, for in so doing it would lose its value to any.

EXPERIMENTAL VIDEO GROUPS

Unlike Open Channel or the Alternate Media Center, the experimental video groups (developing all over the country) are only part-time public advocates. Their primary commitment is to experimentation with portable video. When this aim converges with public access advocacy (and it often does), they readily contribute their considerable skills.

These groups bring to public access a thorough knowledge of the equipment, experience in training laymen, and the sensitivity necessary to work with a variety of people, all somewhat suspicious of the media. For the most part, they live off their work and public access is not a sufficient source of income. Generally, then, the video groups can be counted on for the following kinds of services.

1. Groups like Global Village are professional documentarians serving in a kind of client relationship. You provide the message or choose the subject, and they provide the skilled production services. Global Village grants final approval of the

edited tape to the subject but the editing process itself is completely in their hands. They provide this service free (often as a condition of foundation or grant support) so the tape usually reflects their aesthetics. (Access users should approach such service with caution, since excessive concern with technique can obscure a program's message.)

2. Training is another service offered. But again, because most groups' resources are limited and because the service is usually free, they dictate the terms. Often this means a willingness to train only people serious about video who will continue to make tapes and provide services on their own, long after a particular project is completed. Most groups offer workshops (for a fee), teach classes, or build training into funded projects.

3. Perhaps the most important service, at least at the dawn of a public access project, involves promotion. Video people inspire others with their own enthusiasm about video, helping to generate the initial impetus for public access. Students in particular respond with great zeal, probably because they find excitement in the life-style of the artists. The personal dedication of the video groups has more appeal than a public official's bland advocacy.

In addition, if particular projects appeal to them video groups will contribute what equipment and energy they command. People's Video Theater, for example, is highly committed to using video to aid efforts at self-determination and community control. They have worked hard on a neighborhood project in lower Manhattan to provide basic information on health and other issues.

Public access, as a concept, fits the world view and life style of many video people. They will work hard on the full range of problems—regulatory, financial, technical—to help establish a viable access system. Their experience makes them doubly valuable because many technical and legal people do not understand the production side of access.

This expertise can also have drawbacks. Communities can easily come to depend on a video person or group as the main support of its access system—a potentially troublesome posture. One video person perhaps stated the situation best: "Do we get people to make tapes because *we* like to make tapes or because it will help *them?*" Dorothy Henault of Challenge for Change in Canada offers a similar warning. "People running around with video cameras without the support of [organizers] are going to start stuff they can't finish because they haven't got the skills to do it." A typical example is the person who excites everyone with video's potential. He provides enough information to whet the appetite but not enough to build a project. When he leaves, there is no equipment and no one else to turn to for assistance. All he has done is replace ignorance with frustration.

Of course, most video people do not follow this pattern. The April Coalition is but one exception. It is a loose affiliation of video people, hired by the National Cable Television Association to demonstrate techniques for community origination at its annual convention.

One way to assess a video group is to examine its purpose. Part-time video people, who support themselves by other means, will usually take on only those video projects they are personally interested in. Their enthusiasm can be taken at face value. Sometimes, however, part-time videophiles are hoping primarily to gain experience and build credits. There is nothing wrong with this motive, of course; it does not mean their enthusiasm is not genuine. The point is simply that community groups should not demand or expect too much, depending on them for more than they can contribute in time, equipment, skills, and perhaps even commitment.

Community organizations must strive for continuity, taking care that access to technical and other information, equipment, and fund-raisers survives beyond a video group's local commitment. Training of local people in video skills should also be provided for. With these bases covered, a community organization can welcome and use effectively the considerable resources of video groups.

ACCESS AND LOCAL INSTITUTIONS

For many groups interested in establishing an access center, affiliation with a local institution may help overcome the difficult start-up problems. Lacking recognition, funding, sometimes even a telephone number, the most enthusiastic access supporters may never get their projects off the ground without a recognized local base of support. Institutions such as universities, school systems, libraries, or research centers enjoy legitimacy with funding sources. (The Ford Foundation, for example, rarely considers grant applications from other than large, stable institutions.) In addition, they can often provide overhead expenses until a project generates its own support. Institutions also offer manpower, organizational advice, and a point of identification within the community. These advantages, however, do not come without drawbacks. Institutions imply bureaucracy and control; furthermore, various community segments will have formed attitudes toward them, both positive and negative, that will carry over to new projects associated with them. For example, an access project housed at a university probably will find it hard to get blue-collar workers to use its facilities. These drawbacks can be overcome (blue-collar workers use university hospitals) but only by sensitivity to the problems particular to each institution.

In Port Washington, N.Y., for example, the public library serves as a video center. In its first year of operation, it taught over 450 people how to use a Portapak. Blacks and blue-collar workers unfamiliar with the library were not adequately served, however; future training efforts will push harder to change the library's image and broaden the base of participation.

Port Washington's tapes serve an archival function, chronicling local life and history for residents and future library users. But more important for now are those tapes that trigger responses to current situations.

When a tape of elderly people voicing their frustrations was played back before a local service club, members responded with commitments of time, energy, and funds to begin attacking some of the problems. And in preparing the tape, the old people gained a better understanding of their situation and came to feel a sense of power in communicating it.

Tapes are always played back for the participants and are available for other showings at the library. Port Washington has no cable system, so center director Walter Dale arranges playbacks at service clubs and other civic organizations. Over 7500 people viewed 190 playbacks in the first year, in the project described by Dale as "people explaining themselves to each other creatively."

A library is but one institution and a video center is but one method of taking advantage of institutional leverage. Other institutions and other methods are possible. A "video research" project can be attached to a university. Production facilities can be borrowed regularly from a private company, an equipment dealer, even a church group.

Another possibility is to "piggyback" video on projects like Model Cities or the local Redevelopment Agency. A video component is added to the regular proposal, providing video with a source of funds perhaps not previously available. For example, a proposal to use video for citizen feedback to planners (and vice versa) can be written into a Redevelopment Project, thus generating a ready supply of available equipment.

The problem with piggybacking is that the project or sponsoring institution can administer the funds in a way that severely restricts use of the equipment. (Redevelopment projects, for example, are usually at odds with grass-roots people in the communities they "redevelop," and might restrict their use of the equipment.) The important thing is to decide first what you want to do, and *then* seek out the appropriate institutional arrangement. If you choose the institution first, you may end up doing only what they can help you do, rather than what you want or what is most effective.

The Port Washington project illustrates a noncable use of video. Its chief concern is interpersonal communication and dynamic discussion of local issues and culture. It is easy for a passive viewer to dismiss a program's message on conventional TV. But having traveled to a tape showing, surrounded by his neighbors and face to face with the program's participants, the same viewer can hardly ignore what is said. He may not change his views, but he is more likely to think about them.

Perhaps when cable finally comes to Port Washington, the citizenry will be so experienced and so well prepared that they will discover ways of transferring the intimacy of the Video Center to a cable channel. In any case, the Port Washington Library presents a powerful argument for community video, whether or not your town is cabled. At the very least, when cable comes, you will have a better understanding of how to use it.

WORKING WITH THE CABLE OPERATOR

The easiest, and in the short run, most effective way to begin an access project is to work with (not for) the cable operator. He has the facilities and equipment, the operating funds, and control of channel scheduling needed to produce programs. His cooperation is sometimes the only missing ingredient.

Operators who do cooperate receive excellent public relations (useful when applying for other franchises), good will in the community, perhaps increased subscribership, and the abatement of political pressure. It is a tradeoff whose possibilities operators are only beginning to appreciate.

Reading, Pennsylvania, furnishes the most successful example of operator/community cooperation. This unique project resulted from an agreement between the Alternate Media Center and the American Television and Communications Corporation, owner of Reading's Bucks County Cable TV. The agreement called for one of Alternate Media's top project organizers to live in Reading for four to six weeks, establish an access center, and train local people to use portable video equipment.

ATC provided the physical space, a telephone, and half of the organizer's salary and living expenses. Alternate Media provided the equipment initially, but the project's early success prompted Bucks County Cable to replace the New York equipment with its own two Portapaks and an editing deck.

The Alternate Media organizer, Phyllis Johnson, promoted her free video workshops through a newspaper ad accompanied by a news feature. (Her ads and a set of equipment-use instructions for local citizens are reprinted below.) She reports that the initial response to the story came not from "social militants looking for a forum" but from people "looking for a means of self-expression, or representatives of Reading's many nonprofit organizations desirous of publicity."

YOU CAN BE THERE
Producing Your Own TV Show

That's right . . . Cable Television can be the means of telling the Community your story in your way. You run the show. You can operate the portable equipment and take it anywhere after two or three simple lessons. Learn all about it today and soon your show can be on Cable Television.

FREE WORKSHOPS
NOW FORMING

All You Need Is:
- An Interest In The Community
- A Cause To Talk About
- An Idea To Explore
- You Tell Us

CALL PHYLLIS TODAY

BERKS CABLE CO.
PHONE 376-6341

We're ready for you to watch:

Video Tapestry
produced by

the new
Community Video Workshop

Reading citizens are creating and producing their own programs for cable television **with the camera that goes with the people, where the story is.**

See it on **Cable Channel 5**

Repeated 3 times for viewing convenience

Wednesday	Thursday
1:00 — 2:00 p.m.	10:30 — 11:30 a.m.
9:00 — 10:00 p.m.	3:00 — 4:00 p.m.

This week:

RED LIGHT OR RED TAPE:
> human stories on the scene, of how the fear of traffic affects the lives of residents of Schuylkill Ave. and Ave. A.

ONE SUICIDE A WEEK:
> a discussion of suicide in Berks County and the HELP emergency telephone service, videotaped right in the living room of a Reading citizen.

It could be your story —
> New classes are forming. If you would like to join this free workshop, **two** lessons will start you off.
> **Call 376-6341**

BERKS & SUBURBAN **CABLE CO'S.**

BEFORE YOU GO OUT TO SHOOT

YOU SHOULD KNOW ....

WHO ARE YOU?

You DO NOT represent the Berks TV Cable Co.
You are a private citizen and a member of a community workshop that is open to anyone.

THE EQUIPMENT

POWER

AC Power Pack (converts wall current to direct current)
Connect to External Power In. Turn on power.

Batteries
Take along more battery power than you think you will need.

BP 20 Battery
How to put in back of deck. How to check whether charged.
Duration: 30-45 minutes.
Take one for each tape you intend to shoot.
To Charge: Plug into AC power pack for 3 hours (or more).
REMOVE BATTERY WHEN YOU ARE THROUGH SHOOTING.
If you do not return it fully charged, leave it on Charge,
with a note stating the time you left it.

BP 30 Battery
Connect to External Power In (same place as for AC power).
Duration: about one and a half hours.
To Charge: Connect white plugs.
Use wall current, 8 hours for a full charge.
If you do not return it fully charged, leave it on Charge,
with a note stating the time you left it.

THREADING THE DECK

When putting on reel, position properly on pin.
Follow arrows drawn on deck. When in doubt, check diagram on lid.
HANDLE TAPE CAREFULLY. Do not drag it through the mire of life.
When tape is threaded, DOUBLE CHECK.

Positions on recorder (know what each one is for):
Rewind -- Stop -- Forward -- Fast Forward -- Record.

PLAYBACK

Using Special Monitor
8-10 pin connects deck to monitor.
Set deck on TV position, monitor on VTR position.
Thread tape. Put deck on forward.

Using Ordinary TV
RF unit: convert signal on tape to TV frequency.
Connect RF out to VHF antenna connection.
Set channel selector on Channel 3 or 4.

RECORDING

Use extension cord, unless you are carrying deck on your shoulder.
Connect extension cord tightly and screw in.
Put deck on record position.
Check camera/TV switch (recording/playback).
Tracking control used in playback only.

SOUND

Microphone in Camera: Know its capabilities.
Connecting External Mike: How to keep plug from pulling out.

Handling Mike: Prevent noise.
Do not handle cable connection. If possible, do a test of the
sound in the room before beginning your recording. When others
are going to handle microphone, explain its use.

CAMERA

Vidicon Tube
What are burns? How do they happen?
Try to keep camera in horizontal position when shooting and otherwise.

Lens/Contrast (light level)
Keep in closed position ("C") at all times when not recording.
ALWAYS shoot with higher contrast than you think you want.

> #### Other settings:
> Zoom.
> Focus (most difficult in wide angle, but do the best you can).
>
> #### When are you shooting?
> Turning the camera on and off: two methods.
> Red light on when recording.

WHEN YOU ARE SHOOTING

Begin with 30 seconds of blank on beginning of tape.
Lens in closed position, mike cable plugged in.
Record 30 seconds.
Try to do a test first to be sure heads are clean.
Playback through viewfinder.
Listen through earphone in side-pocket.

Position of Deck When Shooting (IMPORTANT)
What makes the tape wind around the capstan?
When this happens, work it out gently or cut carefully with razor.
Check to be sure heads are still clean.
When you carry deck from place to place, recheck threading.
Place deck in location where it will not be bumped into.
Keep lid tightly closed.

LIGHTING

Bracket
Where will it go?

Poles
How to attach bracket. How to position them safely in room.

Bulbs
Wattage: Know where fuse box is, if wiring is old.
No switch to turn off and on. Unplug when not using.
LET COOL before removing.
Bounce light versus direct light (when and why?).
Keep light out of picture.

HEAD CLEANING

Symptom of dirty heads: sound but no picture.
You don't know till you play back.

Cleaning with Spray
Always do it before taking out deck.

Cleaning with Swabs
If test indicates this is necessary, use gentle horizontal motion.

CHECKING OUT EQUIPMENT

Sign Up on Board
Make pre-arrangement for special requirements:
taking monitor, extension cables for microphones,
extra lights (more than 3).

Sign Sheets When Checking Out Equipment
Sign for everything.
Indicate numbers of equipment and tapes.
Take head cleaner.

When Returning Equipment
Check off everything one by one.
Report all malfunctions, broken bulbs, etc.

LABELING TAPES

After shooting tapes, rewind, and make sure number on reel
corresponds to number on box.
Make up title and put it on label on side of box (THIS IS IMPORTANT).
On front of box, make a record of use the tape has been put to:
date, title, your name, description of tape.

SOUND DUBBING ON PORTAPACK

Turn down volume on monitor after you find place where dub
should begin.
Connect microphone.
Engage sound dub, and place recorder in forward position.
Start dub.

The first trainees came back and brought friends with them. Meetings with poverty groups and classes at neighborhood centers helped bring in other segments of the community, and the workshop grew. After twelve weeks, 60 people had been trained in production with 30 expected to be regular programmers. Phyllis Johnson was succeeded by one of the original members of the workshop, Joseph Masciotti. A Reading native, Masciotti will work full-time coordinating the project, teaching classes, and scheduling use.

Thus far the center offers a one-hour program, "Video Tapestry," repeated four times weekly in different time slots to reach all segments of the audience. It is aired on the local origination channel. The grandfathered cable system is not required to provide access and in fact has no separate access channel.

The Reading project illustrates several important points. No one in Reading asked for or understood cable access, and until Alternate Media arrived there were no local video groups. Yet the project caused great excitement, testifying to the need for organizing and to the great potential effect of a small but skilled access force.

When the cable operator first opened his doors for access, only well-organized civic groups stepped forward to program. When Phyllis Johnson came in, the enthusiasm spread to housewives, blacks, and Spanish-speaking people who finally saw a chance to have their problems honestly reported. To repeat: while the cable operator can lend welcome support to an access center, he is not the man to run it.

Another important example set by Reading is the rich variety of programs that resulted. Local issues were explored, but so were children's poetry, bridge lessons, Spanish culture, and "Projects for a Rainy Day."

Reading illustrates what can be done with the cable operator's good will and warm cooperation; but the Reading experience may not prove to be typical. Cable operators in other communities may do little more than tolerate access and provide meager support; if they do even that, however, they can at least ease political pressures and local groups can build audiences. In Flint, Michigan, for example, Doyle Dugan's black-oriented programming attracts audiences in the black community, but Dugan says he still has to struggle to raise money; the operator offers only limited support.

In sum, cooperation between operator and community can be the least expensive and politically complex approach to access, but you need to keep your eyes open to all the alternatives. One of these is the formation of a public access board.

PUBLIC ACCESS BOARD

The public access board is a device to institutionalize equal treatment for all access users. First proposed by the National Film Board of Canada's Challenge for Change Project, the board is a legal entity responsible for regulation and administration of the access channel. Because creation of the board exceeds FCC provisions, it requires a special showing before the FCC. If the guidelines for access published recently by the NCTA are any indication, cable operator support at the showing may not be difficult to obtain. NCTA recommends that "cable operators should consider the formation of broad-based democratically constituted groups to advise and assist in policy development, promotion and funding for public access."

The access board would likely be set up as a nonprofit community corporation. It would serve as the final arbiter in disputes over scheduling, first-come first-served time allocations, and other questions regarding equal treatment of access users. It also could control access production equipment and funds, serve as an information clearinghouse and access promoter, publicize programs, and promulgate rules for channel use. These various functions (particularly production assistance) could also be parceled out to other groups—a sort of separation of powers to prevent complete control of access by a single body.[11]

The formal access board differs from the informal Reading approach in that, being an institution, its conditions do not vary with changes in cable ownership. In Reading, access partly depends on the good will of the cable operator and the people appointed to run the access center. The access board will survive its individual members, thus offering more public visibility and continuity.

An access board is not at the mercy of the operator's whims and prejudices. Since the board controls the access channel, the operator functions merely as its common carrier, responsible only for delivering a good signal. In fact, it is possible that cable as a whole will someday become legally classified as a common carrier, like telephone and telegraph companies, in which case cable companies will have no programming role. At that time, some version of the access board would likely be adopted in most communities.

But the board approach has its own problems. As with any commission, the method by which access board members are selected is crucial to the board's overall performance. With program monies, equipment, and channel scheduling at stake, access could bog down in a political quagmire. Board members could be elected at large, or by district, or by various constituencies such as unions. Or they could be appointed by the city council or its designated representative. The most important consideration is that all segments of the community agree to the selection formula so that the board has at least some chance of carrying out its task. If the access board simply mirrors the politics of the city council, minorities and other groups will probably be allowed token participation. Their needs will continue to be poorly cared for, and the access system will lose perhaps its most vital constituencies. (Election by district, with restricted campaign spending, seems the most equitable solution.)

But even if the member selection problem is overcome, a more complex problem persists. An access board is a formal, bureaucratic creation. Generally, the more formal a body, the less flexible its policies—and cable access demands great flexibility. *Effective access is usually due to a small energetic group of people working directly with individual citizens and community groups.* An access board could become an intermediary, holding organizers and programmers accountable for their every action. The flamboyance that is often necessary to make access work might be held in check. And it probably would not be long before a public board, subject to the pressures of the ballot box, started imposing such notions as "community standards" or "public benefit" onto a process designed to eliminate such homogenizing elements from the media. Of course it can be argued that with limited resources, standards of allocation and program accountability are necessary. Yet the dangers of over-bureaucratizing access cannot be emphasized enough.

Possible solutions include divesting the board of any programming responsibility or at least having a separate program board for each neighborhood studio. Another approach would be to have the board members appointed on the basis of affiliation —say, one member from the consumers union, one from the welfare rights organization, one from the chamber of commerce. While this eliminates ballot box pressure, it of course raises again the problem of the board's overall balance. A final

[11] An access board was tried in Dale City, Virginia; it is thoroughly documented in N. E. Feldman, *Cable Television: Opportunities and Problems in Local Program Origination*, The Rand Corporation, R-570-FF, September 1970. See Appendix A below.

method is to contract with a nonprofit organization to run the channel until such time as the volume of complaints or the level of discontent warrants termination of the contract.

Because of the danger of politicizing and bureaucratizing access, communities may be better off at the start with less formal methods. However, since the special FCC showing requires the cooperation of the cable operator—best obtained during the franchising process—it may be necessary to include a provision for an access board in the original franchise ordinance. The ordinance could permit the community (or the city council) to invoke the access board provision at its own discretion, thus allowing flexibility in the early period.

The need for an access board partly depends on the cable operator. The following experience demonstrates that in some situations, an access board is the only workable solution.

In Thunder Bay, Ontario, a community organization called Town Talk was programming regularly on the cable. With training and some equipment from Challenge for Change, and flexible scheduling and technical help from the cable operator, Town Talk was able to produce programs on everything from lessons on butchering a side of beef to poetry readings. The District Labor Council, the Consumer Association of Canada, and local artists and musicians all took part in productions. Then trouble began when a large multiple cable system owner bought out Thunder Bay's independent owner.

The new owner considered much of the programming too controversial. Shortly after a live cablecast of a discussion on housing at a public meeting, all live programming was cancelled. Channel time became scarce and a regular schedule was no longer possible. After six months of operation, Town Talk programming ceased.

Despite the rules, a cable operator can influence programming on the access channel. *But with an access board, conditions do not vary with a change in system ownership.* In Thunder Bay, prior to the change in ownership, enthusiasm was high. Even when money was hard to come by, Town Talk supporters managed to scrape up whatever was needed. The citizens even became independent of their trainers, relying on them only for editing help and handling all other production tasks themselves. A Town Talk participant describes the situation after the change in ownership:

> Today in Thunder Bay you can still gain access to cable television. Make an appointment for studio time to tape your program (at last check there was a three-month wait). Arrange your material so it can be said and shown in a 15 by 15 foot area, and hope that the secretaries who have replaced Maclean-Hunter's regular cameramen are as good behind the equipment as they look.

SOME SUGGESTIONS FOR COMMUNITIES

Different access strategies are appropriate for different communities, and choices will vary according to the segment of the community making the decision. In many towns, the system will evolve more by circumstance than choice. As video groups become interested in cable, or Model Cities discovers that funding is available, or the local library decides on its own to imitate Port Washington's successful program, the structure will simply happen. For example, an agreement signed between a Denver-based cable company and a New York City facilitator group determined, at least for the short run, the shape of access in Reading, Pennsylvania, and the project is the most successful in the country.

In fact, Reading's success is ironic in that all of the other examples cited are well planned and supported by foundations or the ubiquitous New York State Council on the Arts, the principal funding source for video in the United States.[12] Reading's access center exists on a $15,000 budget—well below those of foundation-funded projects.

As Reading and Port Washington indicate, a major prerequisite to success is getting the "right people" to run the project. This unfortunately imprecise term seems to mean people capable of working with all segments of the community, who understand and are experienced with video but are less television producers than community organizers. Walter Dale of Port Washington would add that the community organizer must get to know his community well, know its people, their concerns and interests, so that access reflects their needs.

Another common element in successful projects thus far has been that they all began without formal structure. The enthusiasm and energy necessary to launch the projects were free to develop without the constraints of a heavy-handed bureaucracy. If access is treated like a public utility or like a local civic arts commission with a prominent board of civic leaders, its appeal will never drift much beyond the Chamber of Commerce. True, access users collectively need a supportive environment to negotiate adequate franchise provisions, arrange funding, establish good relations with the cable operator, and set up training programs; but individually they need autonomy, the freedom to move without undue bureaucratic constraint. Communities considering an access board would therefore do well to separate administrative and programming responsibilities.

There are other important choices to be made. Who should do the shooting and editing? People with something to say may not be interested in video, or may be time-poor, or may prefer to draw on professional skills. These people are out to attract an audience with a good program and will need a facilitator service such as Open Channel or Alternate Media. Other people will want to do their own production work, and will need only equipment and training or the kind of service that the Reading Video Center can provide. For these people, conceiving, shooting, editing, and presenting their own programs will teach them a great deal about the myth of objectivity in mass media and sensitize them to media manipulation.

Still another approach is to ignore cable (except as a secondary interest) and emphasize organizing and involving people in local problems. As Gwynne Basen, a video organizer for Challenge for Change, puts it, "Either you're there to start a video group or you're there to help an organizer build a citizen's committee around whatever issue they have to work with." She adds that it is important to know when to phase video out.

In this context, video is a catalyst that may no longer be necessary and can be phased out once effective community action on the problem has been organized.

---

[12] Each state has its own State Council on the Arts, but only New York's is really active. Two years ago, for example, the California State Council's overall budget was under $200,000, New York's over $18 million.

Gwynne Basen used video to help form a citizens group to combat noise pollution caused by trucks barrelling down their street. Once the group was formed, she and video bowed out in favor of a local organizer. She and other organizers feel that this kind of follow-up to taping is crucial, arguing that "to make programs for cable about people's problems without helping them deal with those problems is just as bad as commercial TV."

Different approaches serve different goals and values and reflect different conclusions about our communication needs. In thinking about which access model to establish or what kind of service to offer, you must first determine what you need or are able to accomplish. If you are lucky, funds will be available for all approaches, but such luxury is not likely and hard political battles seem inevitable.

V. PROGRAMMING COSTS AND EQUIPMENT

While access program costs depend on several factors, equipment is one of the major budget items. The kind of equipment you choose affects the type of program you can produce. A Portapak, for example, offers great mobility but has technical limitations. Half-inch signals can cause problems in cable transmission, and small portable color cameras and recorders are just being introduced. The studio, on the other hand, offers adequate technical quality and a complete color capacity but severely restricts field recording possibilities. Three possible configurations of equipment follow:

1. *The studio* consists of tripod-mounted cameras, lights, microphones, a switcher, video monitors, and a videotape recorder (VTR), usually one-inch, housed in a soundproof, air-conditioned room, well supplied with electrical outlets. The switcher, usually housed in a separate control room, selects among incoming camera signals previewed on control panel monitors.

2. *A multicamera set-up or portable studio* is basically the same but uses small cameras (either hand-held or small tripod studio cameras), smaller portable lighting facilities, and a more limited switcher/control panel. Either one-inch or ½-inch VTRs can be used.

3. *A simple Portapak/microphone* set-up consists of one or more hand-held unconnected camera/recorders.

THE STUDIO

Of the three configurations, the studio can deliver the best technical quality because it enables complete control of lighting and acoustics and careful placement of camera and talent. The studio is free from obtrusive natural sounds, variations in sunlight, and troublesome onlookers.

With a switcher, the director can cut between different camera angles and positions, going from a close-up to a wide shot—giving visual interest and a sense of pacing to liven up even the most pedestrian discussion. Multicamera flexibility usually eliminates time-consuming editing and saves costly tape. Rather than shooting six tapes to produce a single edited tape (referred to as a 6-to-1 shooting ratio), you shoot one tape, varying shots with the switcher as you go along. Of course this method can't eliminate the dull moments or change the actual sequence of discussion but it saves post-production (after the shooting) time.

A studio is best used for straightforward discussion or interview programs where mood and setting are not terribly important. For example, a program offering basic consumer information is easier to tape in a studio where everything is already set up for production. And graphics, slides, even Portapak tapes shot out in the community can be integrated into a studio program.

Plays, dance productions, and other performances usually require studio-quality lighting control and camera mixing. Such programs demand substantial pretaping preparation and scripting and an experienced crew and director.

The studio does not serve all programming needs, however. Above all, it gives up flexibility in exchange for technical quality. The open-ended, spontaneous neighborhood tapes of the Alternate Media Center are impossible to duplicate in the confining studio setting. Many people are ill at ease before the lights and television cameras. Removed from their natural surroundings, an entire dimension of their lives is stripped away and the subtleties of an unrehearsed tape are sacrificed for technical control. To visualize the difference, imagine seeing a factory worker actually at work on an assembly line as he describes the deadly boredom of his job, and then imagine the same man brought to a studio to voice his message against the same backdrop used by the local Dick Cavett or Johnny Carson. Seen on the assembly line, the worker dramatizes a serious social problem; in the studio, he risks coming off as a cocktail-hour complainer.

Another problem with studio programming is that too often it tries to mimic broadcasting. Without the technical expertise, the same equipment, or the on-camera talent, the results are often disastrous.

Studios have their place, but their inherent rigidity may unnecessarily limit access producers. If possible, more intimate or spontaneous programs should be taped outside the studio. High-quality mobile units are one answer but they are impractical for most access users. Portable video, described in the next section, seems the best complement to a studio facility.

Basic one-inch studio costs are shown on page 44 in a table reproduced from an Urban Institute publication.

MULTICAMERA, PORTABLE SET-UP

The portable version of a studio includes a VTR, two to four cameras and tripods, a small switcher with preview monitors (to indicate what each camera is seeing at that time), an intercom between director and camera operators, portable lights, microphones, and a sound mixer. If hand-held cameras are used and the recorder is light, this "studio" can be taken anywhere.

Used primarily by Open Channel, this system provides the same camera-mixing potential as that of the studio with similar savings in editing time and tape cost. As with a studio, the multicamera set-up is sometimes too imposing for intimate taping situations or too cumbersome for more dynamic events such as a fair or demonstration. Used primarily on location, this approach does not offer the lighting and acoustical control of a studio. In addition, the portable studio has to be completely set up and taken down again—lights, cameras, cables, and microphones—for each

Basic Equipment for Cablecasting

Equipment	Black and White (B/W)		Color		
	Basic	Average	Basic	Average	Full
Figure		II	III	IV	V, VI
Cameras	2	2	2	2	3
Tripods	x	x	x	x	x
dollies		x	x	x	x
zoom lens	x	x	x	x	x
headsets	x	x	x	x	x
Control					
switcher	x	x	x	x	x
special effects	x	x	x	x	x
Film Chain		x		x	x
Audiotape player		x		x	x
Monitors, Video	1	4	3	4	5
Waveform Monitor				x	x
Microphones	2	3	2	3	4
Mixer, audio	x	x	x	x	x
Audio Monitor		x	x	x	x
Video Recorders	1	1	1	2	2
tape size	1/2"	1"	1"	1"	1"
Lights	x	x	x	x	x
Approximate Cost	$10,000	$20,000	$35,000	$50,000	$80,000

SOURCE: Charles Tate (ed.), *Cable Television in the Cities,* The Urban Institute, Washington, D.C., Copyright 1971, p. 43. Reprinted by permission.

taping situation. This task, plus the time pressure of finishing the work and clearing out of busy facilities, means a large crew is often needed. One Open Channel tape, a pulsating 2½-hour black church service, took a professional crew of 20 an entire day to produce!

The multicamera approach is best suited for planned events too large for a studio, such as concerts or rallies. It is essentially an inexpensive way to bring limited studio capacity to the community when the community can't or won't come to the studio. Its chief virtue is that it reduces the need for editing, which, as the next section indicates, can be troublesome with half-inch.

A basic small multicamera system follows:

3 small Vidicon cameras with tripods[a]	$2000
1 switcher/special effects generator	500
1 ½-inch recorder[b]	800
lighting kit	350
assorted cables	50
3 mikes..	150
mike mixer	150
(videotape—½ inch, avg price $16/½ hr)	
Total	$4000 (+ tape)

  [a] Tripod studio cameras are more expensive—on the order of $2000 each.

  [b] A one-inch recorder can be used just as easily, provided cameras are adaptable.

PORTAPAKS

A Portapak with an outside microphone clearly affords the cheapest and easiest production approach. No lights are required in outdoor situations, and indoors a new low-light camera tube, the Tivicon Ⓡ, eliminates the need for lights in many cases (although lights are recommended where convenient).

The Portapak is best suited for large, unrehearsed, or spread-out events, intimate or informal situations, or simply unobtrusive documentation, such as a tape of a classroom. A two-person crew is enough for most situations. (Since a Portapak contains an internal mike, theoretically one person is sufficient. In practice, however, tapes with sound shot by a one-man crew are uniformly inferior.)

The interview is the commonest Portapak program, and its key element is the technique of the interviewer. He must decide just how manipulative to be: whether to press hard for succinct responses to a predetermined line of questioning (it saves tape), to let the subject go on at his own natural pace about randomly selected subjects, or to become an active participant in the interview—making himself as vulnerable as his subject. The choice depends on the kind of tape desired, but fair treatment of the subject should also be a factor. You should take time to make the subject feel comfortable with both interviewer and equipment, perhaps by playing back a short piece (viewed through the Portapak eyepiece), before the interview really gets rolling. Shooting time should not be measured, nor the subject battered into making pat statements. Network-style manipulation of the interviewee usually gets stock responses that evidence little thought or insight. A more patient approach may use up more tape but the end result will be worth the extra effort—provided editing is possible.

Most Portapak productions involve more shooting than the final tape can use. In single-scene interviews or several-scene documentaries, at least some editing is usually warranted. Editing can present technical problems. The best method of half-inch editing is to electronically transfer or copy (dub), from one machine to another, segments of ½ original tapes in a desired order onto a master tape on a second VTR. The second VTR is specially equipped to provide edits free of technical disturbance; in half-inch it costs between $1100 and $1500. A good one-inch editing machine costs about $5000 (you can transfer from half-inch to one-inch, and recently even to two-inch broadcast standard). The problem is that with each copy (generation), signal quality is slightly degraded and the picture may suffer in cable transmission. This is one reason Open Channel adopted its multicamera approach and Alternate Media simply did not edit its early tapes. Half-inch editing sometimes causes sound and visual disturbances at the juncture of two segments, but this situation is improving as new editing machines and processing equipment are developed.

Since electronic transfer is the best editing method, you should be sure your cable system is adequately equipped to transmit second-generation tape. (Electronic editing means copying segments. The copy—a so-called second-generation tape—loses some of the quality of the original taped segments. A copy of a copy—a third-generation tape—loses a bit more, and so on.) With proper support gear—e.g., a good processing amplifier—and adequate maintenance, cablecasting of second-generation tapes should achieve acceptable results. Editing on one-inch increases the chances for success but also increases the cost. If possible, you should use a one-inch editor, although second-generation problems are possible in any case.

Other editing methods are available but each has its own limitations. Scripting to allow editing in the camera usually results in stilted productions, and the "edits" rarely escape some technical disturbance. Physically splicing the tape permits cablecasting of first-generation tapes, thus insuring decent image quality for reception, but again the edits are not absolutely clean. Because splicing damages the original tapes, their reuse is limited (recording on spliced tape is not recommended for proper video-head care).

If reception is not a problem, electronic editing is best; otherwise, splicing is. (Reception should be less of a problem as the editing and support gear improves and more is known about the problems, so try it if at all possible.) Finally, in some situations when skilled help is available, the multicamera operation described above can be used to limit or replace editing.

The Reading project equipment budget appears below. The prices are well below list. Portapaks list at $1650 each and the editor at $1095. Thirty-minute reels of ½-inch tape list for about $20 but few people ever pay more than $15 per tape and many pay as low as $11. The Reading budget provided for only one full-time paid employee. The cable operator provided much-needed technical assistance and repairs. (It may be possible to arrange this on a cost-plus-fee basis, since the operator is equipped for maintenance.) Each access center should seek a repair alternative to the hardware distributors, who charge $20 an hour labor. Jobs such as changing the heads can be done easily with little technical training, and you will save $30 dollars a crack (minimum labor charge in most places). Many video groups have some technical skills, and will usually do simple repair work for $10 an hour labor or less.

The budget for the Port Washington Video Center is an example of a more extensive operation. The State Council grant pays for a full-time director, some part-time staff, and the following equipment: two Portapaks, two decks including an editor, four television monitors, microphones, lights, long-life battery packs, 400 to 500 hours of videotape, and a little-used video switcher that allows several camera inputs to a single recorder. At least two-thirds of the budget is for personnel, and the library provides the space and some operating expenses. Walter Dale, project director, recommends an additional $5000 annually for maintenance and repair. Funded by the New York State Council on the Arts with $45,000 for 18 months, the center expects to become self-supporting soon after the grant expires (through com-

BILL OF MATERIAL

Ordered on PO 9766

System Name **Berks TV Cable Company**   Capital Appropriation Request No. **49-21**

Name of Vendor	Acct'g. Code	Description	Part Number	Quantity	Unit Price	Material In Inventory Total Cost	Material to be purchased Total Cost
Technisphere		Sony Portopack Deck	3400	2	1270.00		2540.00
		Electrovoice microphone	635AKG	2	69.00		138.00
		Sony 8" monitor		2	165.00		330.00
		12' camera extension cable		1	50.00		50.00
		16' camera extension cable		1	33.15		33.15
		Microphone extension cables		2	4.50		9.00
		½" Video tape half hour reels		48	12.00		576.00
		RF Units		2	42.00		84.00
		Sony Editing Deck	3650	1	845.00		845.00
		Sony Cable Connection set for editing		2	10.80		21.60
		Lowell Portable lighting kit		1	200.00		200.00
		(6 bulbs - 2 cords - 2 stands - 6 brackets- carrying case)					
		This quote was received from					
		Jack Goldman					
		Technisphere					
		141 Lexington Ave.					
		New York, N. Y.					
		Ph 212-684-3136					
		Quote requested by Phyllis Johnson					
		Alternate Media Center NY University					

Total Material On Hand Cost

Total Material For Purchase Cost

Total Material Cost   4826.75

ATC Form 403 8/71

Note: Refer to the latest systems P.O. File to find unit prices for items listed.

munity and library funds). Libraries in San Francisco and elsewhere are beginning to attempt video centers of their own, and although Arts Council money is not available to them, they still look to Port Washington as a model.

In its first year of operation, the Video Center produced 600 hours of programming. The director determined which tapes to preserve and which to erase and reuse. The 60 local people trained in editing produced 200 hours of edited tapes.

PRODUCTION EXAMPLES AND COST

The cost of setting up a full-size access system with facilities available to an entire community has been described. But there is another perspective, that of the individual user or group interested in producing a tape. In many communities with limited access demand, the cable system will be completely responsible for access, in which case groups may well want to seek outside production help. Communities that have independent access facilities may encounter problems of equipment availability or personnel flexibility, again raising the need for outside help.

Outside help comes by way of film students, professional video groups, or film professionals willing to work for free or far below scale. Equipment can be borrowed as needed from schools and other institutions, or rented from a cable-system hardware distributor or production groups. Probably the cheapest and most efficient rental approach is to buy production services from a crew that has its own equipment.

The buy-versus-rent problem is addressed neatly by David Othmer in the special issue of the *Yale Review of Law and Social Action,* "The Cable Fable." In his article, "Portable, Half-Inch Production: A Note on Costs," Othmer points out that while rental arrangements vary substantially, a commercial rule of thumb is available:

    1 day rental—5% of equipment value
    1 week rental—15%
    1 month rental—30%

Complete studios generally rent for a flat fee, often as low as $25 an hour but sometimes reaching as high as $100 or $150. Thus the decision to rent or buy equipment depends on production needs and the long-run estimate of rental costs compared with purchase prices. The tradeoff, according to Othmer, is easily evaluated. "If, for example, one expects to do a reasonable amount of shooting, but little electronic editing ... buy the Portapack, but rent the second deck."

Two other factors enter this decision. First, video equipment continually improves and older models become obsolete, thus tempering the value of purchase. (This will always be true, of course, but cable technology has not settled down enough to justify investment unless extensive use is planned.) Second, rental rates from video groups are perhaps 50 percent of commercial distributors' rates, and perhaps lower if a group is hired to produce the tape.

Production offers a confusing number of variables in addition to those concerning equipment. How many crew members are needed for how many days? How much tape is required and how much producer preparation before shooting begins? Should you produce, hire a producer, or seek volunteer help? (For a helpful model see Appendix A, which presents Dale City Television's basic operating procedures.)

The decisions depend upon resources, quality desired, and other factors that each group must assess. Fortunately, some guidance is possible. Minimum crew size, for almost all situations, includes a separate camera operator and someone to handle sound. The "producer" can fill either role or be a third crew member. Situations such as large events require more than one crew. Student video crews will charge between $2 and $5 an hour for camera operators. Sound people demand closer to the lower end. Professional video people charge more, $10 an hour and up. But the promise of future work, a particularly interesting project, or an impoverished group will bring the rates down. Community groups can train their own people to serve as crew, but while it is easy to learn to operate ½-inch equipment, using it well requires much skill and practice. Sound people, however, should be easier to train (although sound quality is the greatest drawback on most tapes). Producer/director types—at $50 to $125 dollars a day—are probably too expensive to hire, but can often be counted on to volunteer at least some assistance. Most community groups will assume responsibility for their own ½-inch productions, relying on more experienced people for advice.

Othmer suggests one day of organization and planning for each day of shooting, and this is certainly wise for complex productions. Simple interviews are often done with little advance planning. But in general, locations must be scouted, contact made with crew and participants (particularly the latter, to establish trust), and thought given to the intent or hope of the tape, its style, and content.

Despite the problems and decisions, ½-inch production for cable is possible. A good example is "The Elders," 12 shows 40 to 60 minutes long, produced by David Othmer for the entertainment and education of the over-65 community in New York City. Taped in day-care centers and communities around Manhattan by a two-man N.Y.U. student crew, the program "demonstrates what is going on in the elderly community." It is not a series about the problems of people over 65; rather, it portrays the variety in their lives.

The total cost of the 12 shows, excluding producer/director fees and using physical splices to edit, was $2800 or $233 per show for a 50-minute program. The tabulation at the top of page 50 is a cost breakdown.

Compare Othmer's budget with one from an award-winning, high-quality program from Continental Cablevision's Findlay, Ohio system. Produced with a staff of 12 (not all full-time) and a $100,000 mobile unit, "Just Imagine" is a noncommercial "bedtime story for preschoolers and kindergartners."

The stories are told through narration to children selected as "listeners" for each show. Guest storytellers include a local magician, a puppeteer, and teachers. Locations include the public library (the source of story materials), the local playhouse, nursery schools, children's bookstores and even a turkey farm for the Thanksgiving story.

The 15-minute "Just Imagine" programs are aired daily. According to program director Jeff Jones, "In order to keep costs down and also to facilitate ease in taping schedules, we found it necessary to produce three programs per week and play two of them twice. All three programs are taped on Saturdays, if possible." The cost

```
DAILY PRODUCTION EXPENSE

    Depreciation allocation:  $16.75/hr × 1 hr ......   $ 16.75
    Talent/producer ..................................     16.00
    Personnel:
      2 cameramen
      1 floorman
      1 technical director
      1 videotape operator
      5 × $2.00/hr × 1 hr ...............   $10.00
      1 director .......................      8.00
         Total personnel ...........................   $ 18.00
    Graphics .........................................     10.00
         Total per program ..........................   $ 60.75

MONTHLY EXPENSE

    Production:  3 programs weekly × 4.3 weeks ......   $783.70
    Playback .........................................     11.00
    Personnel:
      1 technical director
      1 film man
      1 videotape operator
      3 × $2.00/hr × 1/4 hr × 4.3 weeks .............   $  6.45
    Advertising and promotion ........................     30.00
         Total monthly expense .......................   $831.15
```

breakdown below is based on the production of three programs per week (figures supplied by CTV-Three). A 15-minute show costs $60.75 to produce, while the monthly costs are $831.15.

```
              PRODUCTION BUDGET FOR "THE ELDERS"


Equipment rental:      13 days shooting,
                       8 days editing deck,
                       2 days viewing equipment .........   $ 800.00

Videotape:             36 half-hours, 12 hours
                       (12 hours for copying
                       originals) ......................     761.00

Camera people:         36 man-days at $30/day ...........    1080.00

Transportation:        car mileage, taxis, buses,
                       subways, etc. ...................      50.00

12 empty reels:        at $1.85 per reel ...............      22.20

Promotion:             Flyers, postage, etc. ............     50.00

Telephone use (est.) ...................................      10.00

Secretarial time (est.) ................................      40.00

    Total ..............................................   $2813.20
```

Based on the program samples offered, production costs can run as high as $250 an hour. At even half this rate, many communities will be hard-pressed to pay for more than a few original hours a week. Program sharing with neighboring communities therefore can be an important step in building community origination.

Neighboring towns often have common problems, and the cost of a single program can thus be amortized over a number of showings in a number of communities.

Exchanging programs by hand (as opposed to interconnection) requires a compatible videotape format to play a neighbor's tape (remember that one-inch machines are not standardized), so new systems should investigate what hardware is available locally and select accordingly.

In addition, a regional processing/editing center for half-inch tape, making available high-quality one-inch editing machines and sophisticated processing equipment, would greatly improve the utility of portable equipment for cable production. The center could also transfer one-inch tapes from one format to another.

In short, the question of program equipment cost must also be looked at from the perspective of an entire region. Compatible hardware, pooled facilities, and shared programs, ideas, and experiences will reduce the per-program cost for a single community and offer greater technical quality.

TRAINING

Learning to use video equipment is neither as hard as it looks nor as easy as everyone says. While almost anyone can learn to operate the equipment in about twenty minutes, it takes experience to produce a technically good tape. When Alternate Media covered a three-day meeting, they tried to train local people to do the taping, but soon realized it was too difficult. And in this case it was more important for the tapes to be seen than for laymen to produce them, so AMC staff members shot that situation.

This raises the question of whether you should have special camera operators or whether everyone should learn how to shoot his own tape. My opinion is that everyone who wants to learn should be enabled to, but that no one should be pushed into it. At least a few people, however, should become very proficient camera operators and editors so that every group will have access to skilled help. These people can also become a nucleus of new teachers, passing on to others what they so recently learned.

Small group workshops are an effective training method. An experienced instructor quickly teaches three to five people the basics of equipment operation and sends them off with a tape to shoot whatever they wish. When they return, the tape is played back and critiqued by the trainer. This immediate hands-on experience is worth more than six hours of lecturing on video technique. Once the tape has been viewed, the trainer can talk in more detail about proper use of the zoom, tips on holding focus, and proper lighting backdrop. While he talks, the trainees can tape him and feed the tape live into a monitor for the trainer to critique as it happens. This little exercise in instant feedback is extremely useful, particularly if it follows a period of unstructured experimentation.

One lesson too often ignored is proper mike placement and use. Many tapes are ruined by poor sound, a problem that a 30-minute lesson could eliminate. Good sound for video requires more concentration than skill. The person holding the mike should be conscious of arm movement, background noise, and mike direction. In many interviews, the mike holder points only at the interviewee, with the result that his own questions are never heard.

Lessons in lighting and editing should be given before tape production begins. The important editing lesson need not include technical training at this stage, but the limits of the technology and the variety of styles available should be explained. Editing offers great latitude for individual expression, but a good tape-maker has some idea of how he will edit as he is shooting.

Finally, proper care of the equipment—cleaning heads, checking plugs and connections, learning how to replace fuses, and an orderly process of tape and equipment checkout—should be emphasized throughout so that each user will receive equipment in workable order. I remember doing a fascinating interview with a man who lived in a cardboard house. I rushed home to play back the tape, only to discover that the wiring on the mike connector had come slightly undone beneath the shield—a routine occurrence but one that careful checking can catch so that valuable tapes are not ruined.

VI. SPREADING THE WORD

DOES AUDIENCE SIZE MATTER?

> *Idealist:* "It'll be fantastic. Television will be as accessible as a fountain pen. Everyone will make programs about everything and no one will censor or interfere in any way."

> *Cynic:* "Yes, but with everybody and anybody producing, nobody will be watching."

This slightly caricatured conversation, repeated everywhere public access is discussed, finds critics arguing that without an audience, public access is pointless. This critical view assumes first that audience size is important, and second that public access cannot compete with other program sources for audience attention (thus far an unsubstantiated claim).

The first assumption is less a matter of fact than of attitude. As pointed out throughout this report, one access strategy is to involve as many people as possible in production, emphasizing participation and identity-building over audience share. While this method may result in a slow but continuous increase in audience, that is not the immediate goal. Another response to the critic's first assumption is that the ability to appeal to tiny audiences lies at the heart of cable's potential. As Pierre Juneau, Chairman of the Canadian Radio and Television Commission, put it,

> It is a paradox: cable programming developed because of the inability of [broadcast] television to contain and cope alone with the new needs of an evolving society. Now its very early attempts to adapt to these needs are judged by the standards of the sector which could not deal with them.

In short, one answer to the cry that nobody watches is, "So what?"

For some access producers, this response is too glib. Their reason for programming is precisely to reach an audience—they want to be heard. The allegation that access cannot compete for audiences troubles these producers. Even if the difference is between no audience and an audience of 3 percent (which may be large, for access), the difference is vital. For people dealing with social issues or appealing to their neighbors on other levels, audience is a crucial concern. As this report has frequently reiterated, while process video and individual expression are important, access will be a luxury few communities can afford unless it enables socially involved groups to reach an audience.

AUDIENCE SURVEYS

Polling techniques make audience attention primarily a numerical issue, usually ignoring program impact on those who did watch. Less formal methods, such as word of mouth in small towns, may provide all the feedback needed.

Building audience size has two components: the complex problem of feedback and promotion. People can't watch even the best program if they don't know when it's on. Producers don't know how many people are watching without some contact with the audience. In commercial television, audience size is measured by A. C. Nielsen & Co., which attaches a box to TV sets in selected homes to record program choice. Access producers must develop their own feedback methods.

When two-way cable finally emerges, public access will have direct feedback on audience numbers. Two-way will permit immediate electronic polling of the entire subscribership instead of a random sample. Producers will receive feedback on all programs rather than just a few, and the audience response process will be cheaper. Until two-way is fully operative, audience feedback can be obtained through telephone polls, by asking the audience to call in at show's end, and by sending out questionnaires or interviewers to find out if people watch the access channel and if so, what programs they watch. The various feedback methods require skilled help to develop the sampling techniques and evaluate the results, but this kind of skill is readily available at universities, sometimes even on the undergraduate level. If help is not available, a more general appraisal may suffice (such as tabulating viewer call-ins).

PROGRAM PROMOTION

Promotion is any scheme, however wild, to bring your message to people's attention—ads on park benches, marching bands, mass leaflet campaigns, and program advertisement coupled with other mailings such as the public utilities bill or an Easter Seals drive.

Promotion can contribute more to audience size than can the quality of the program itself. It can also become more expensive than the actual production. If each access producer had to pay for the newspaper ads, subscriber mailings, and ads on other cable channels or local radio, many couldn't afford to produce any programs. Fortunately, there are several ways to pool promotion; the first three are recommended by the National Cable Television Association to its membership (see Appendix B):

1. Cable operators can include access program schedules and promotional material with their monthly billing to subscribers.

2. Nonbroadcast channel time can be used to list forthcoming access programs either continuously on an empty channel or as spots on utilized channels.

3. An access board or other publicly constituted group can be formed to promote the access channel in general and to help advertise programs.

4. Access program schedules should be regularly listed in local newspapers.

5. Local radio (cable radio, also) should list each day's programming at a regular time, perhaps once after a morning newscast and once again after the evening newscast.

6.   Subscriber lists, usually available to commercial customers, can be obtained from the cable operator for mailings between billings.

7.   Promotion of channel use will draw local attention to the programs.

8.   The following innovative approach was tried successfully on a Canadian cable system. A list of more than 100 tapes produced by local community groups was sent to all cable subscribers, who were asked to phone in the numbers of the tapes they would like to see. More than 1500 requests were received within four days. Each requested tape was aired, with the most popular ones shown several times at prime-time hours. The playback schedule was displayed daily on the access channel. This method obtained audience feedback and promoted programs as well.

Other methods of promoting regular programs include showings to various organizations, schools, or selected audiences. A private showing of a particularly good sample program may encourage regular cable viewing.

Another approach is to involve local celebrities or large numbers of local people in a program. A sports figure providing health information might catch the viewer's eye; so might taping a program on the town green during lunch and informing onlookers that they can see themselves on the cable. These methods assume that the greatest problem is getting viewers to start tuning in. It is to be hoped the program will be interesting enough to retain at least some viewers without having to resort to gimmicks.

Promotion is most successful for regular programs. Specials or one-time programs require splashier and hence more expensive advertising, although reserving a particular time spot for specials may provide some carry-over of viewers from week to week.

Promotion is most easily done by organizations, who are the primary users of access anyway. (Individual users rarely have the resources or the ongoing need to take advantage of access.) A group can encourage its members and their contacts to watch the program, and each organization has a natural constituency outside its actual membership that proper promotion can motivate to tune in. For example, a consumer program probably will attract co-op members and ecology groups. Member lists from related groups should prove useful in building audience support, and this kind of focused advertising is much less expensive than mailing to all cable subscribers.

VII. FINAL SUMMARY SUGGESTIONS

The FCC has issued minimal regulations for access channel administration and individual use. But for access use to extend beyond relatively well-established groups to people and groups active in community organizing and activities—for the progressive sentiment to become a grass roots reality—procedures, funds, and facilities for community-wide use must be developed.

PHASE I: EDUCATION AND ORGANIZATION

The first tasks are primarily political. Local resources such as video groups and facilities, local organizers, neighborhood associations, consumer groups, and legal and technical experts, must be located and tapped to help begin the organizing work immediately whether or not cable service already exists. *Other active groups and organizations working on other issues must be informed about cable and goals must be established.* (Video projects can help both in demonstrating cable and focusing on goals.) If the franchise has not yet been awarded, ownership is an important consideration. If the local franchise ordinance has not been drafted, involvement in that process is essential.

PHASE II: FRANCHISING AND PREOPERATIONS

With education and organization under way, politicking at the franchise authority level becomes paramount. Funding and equipment provisions should be included in the franchise ordinance, or arranged separately with the municipal authority or current or prospective owners. Channel administration options (e.g., a public access board) should also be included even though they may not be immediately implemented. (An informal approach to administration is recommended at the outset.) Neighborhood needs, including equipment and operating expenses, should be included in franchise recommendations. It is most important, however, to secure the political and technical backing to support a franchise and a structure (e.g., an access board) responsive to community demands.

PHASE III: OPERATIONS

The next phase of decisions deals with the need to help access users reach an audience. In other words, after the organizing and political issues are well in hand, including a solid case for the FCC certification process, production-oriented problems can be addressed. You can now get on with scheduling and promotion (the former requires certification if dealt with separately from overall channel administration), equipment choice, finding the right personnel, setting up training programs and attracting users.

In summary, then, the most vital access tasks have nothing directly to do with television, but are concerned with organizing and building a political consensus. Once the proper franchise framework has been established, including channel administration, funding, and facilities, you can begin the extremely difficult basic tasks of attracting and training users, and scheduling and promoting programs. But without organizing and politicking for the best locally determined access environment, production questions are matters for idle speculation.

Finally, if your system is already operating and no support for community programming (not cable-operator programming on behalf of the community) is forthcoming, don't throw your hands up and say, "It's too late." While the task is far more difficult than that facing a not yet cabled community, local pressure can be persuasive. First, negotiate with the operator. If that doesn't work, lobby with municipal officials to enforce or, if necessary, amend the franchise. Find a lawyer and see if legal action is possible under state or federal regulations. Take well-documented examples of support in other communities and of the lack in your own to local media, citizen groups, the city council, state regulators, and the FCC. Tell people what they are missing. Compile detailed lists of all would-be access users and the range of programs they could offer. Make your case known statewide and at the company's national office—negative publicity for companies pursuing other franchises can be a powerful bargaining instrument.

It is to be hoped that such negative tactics will prove unnecessary. Most cable operators support at least the concept of community origination. When conflicts arise, the principal task is to convince both the operator and local citizens that public access will flourish best as a cooperative venture between them. Both partners have a strong stake in the outcome, for the success of public access can strengthen cable television's role in the community at the same time it serves vital citizen interests.

Appendix A*

BASIC OPERATING PROCEDURES FOR
DALE CITY TELEVISION[1]

DEPARTMENT DIRECTORS

Program Director

The Program Director will have complete responsibility over the entire systems operation. All department heads will report directly to him. The Program Director will be responsible in assigning Staff Directors, Producers, Staff Announcers and Assistant Directors for each telecast. All program ideas and policy questions are the direct responsibility of the Program Director. Decisions concerning program policy and station operation are the direct responsibility of the Program Director. He will report directly to the Advisory Board.

Director of Public Relations

The Director of Public Relations will be the official public spokesman for the system. This department will be responsible for supplying the mass media with current information concerning this system. Weekly program logs will be supplied to the local papers from this department. All PR information is to be approved by the Program Director through the Public Relations Office. Anyone is free to submit promotional or public relation ideas; however, they must first be approved by the Public Relations Department. In addition, the Public Relations Director will be responsible for supplying promo-

tional information for particular telecasts and assisting Producers and Directors in the writing and placing of promotional and public relations information.

Directors of Sports, Women's Programming, News, and Public Affairs

The Directors of these departments are required to keep on hand a constant approved backlog of program ideas. They will be responsible for producing their own programs, supplying formats and talent, and organizing each show in such a manner that it is ready for airing when it comes into the studio.

Director of Graphics

The Graphics Director will have the responsibility of producing a complete set of graphics as required for each telecast. In addition, a graphics person will be on hand for each production. Graphic requirements will be submitted to the Graphics Director at least one week prior to a telecast.

Scene Designer

The Scene Designer will be responsible for designing and preparing each set for each telecast. Scene requirements are to be submitted to the Scene Designer at least one week before the proposed telecast. Additionally, the Scene Designer will be responsible for all props for each telecast.

[1]The rather elaborate operating procedures devised for Dale City Television may provide a useful basis for community TV in many other areas as well. For that reason, this description, based on one prepared by David J. Touch, Program Director for Dale City Television, is presented here.

* This appendix is reproduced verbatim from Feldman, op. cit.

Producers

Anyone in the system is encouraged to independently produce programs, under the following guidelines:

1. A program idea must be submitted first to the Program Director and then approved by the Advisory Board.
2. The Producer is required to submit script outlines, budget, story boards, final script, and all production at least one week before the proposed production.
3. The Producer is required to assemble his/her own crew with the assistance of the Production Supervisor.
4. The Producer will be required to supply to the Chief Engineer all technical requirements for the production.
5. The resources of the entire system will be at the disposal of the Producer for a particular program.

PRODUCTION STAFF

Announcers and Talent

As scheduling permits Staff Announcers are asked to be on the set at least one-half hour prior to air time. Coats and ties are required for men, dresses or suits are required for female participants. Sports clothes, etc., are not authorized for on-camera work.

Engineering and Chief Engineer

The Chief Engineer is solely responsible for the system's entire engineering and engineering personnel requirements. He will schedule for each taping a video engineer and audio man. He will be required to supply information on power requirements, camera and audio operation, including mike positioning, to the Producer and/or Director for a program. Additionally, he will be responsible for the setting-up of cameras and the master control area. He will be responsible for maintaining all electronic equipment and keeping it in operating order, and report directly to the Program Director any technical difficulties occurring during any taping session. He will hold primary instructional sessions in television engineering for both cameramen and audio men. The Chief Engineer will report directly to the Program Director.

Production Supervisor

The primary responsibility of the Production Supervisor is to gather each full crew for a particular session. This includes cameramen, floor audio, lighting, grips, and floor manager. Each individual will know in advance what he is expected to do for a particular taping session. It is the responsibility of the Production Manager to see that his crew stays for the entire session and assists in tearing down the equipment in the studio area after the production. In addition, he will be responsible for familiarizing new people with the systems operations along with formally introducing new people to the television staff. He will be additionally responsible for the scheduling of teaching sessions for both production and engineering, along with teaching primary production techniques to new members. The Director of this section will report directly to the Program Director.

Production Staff

The Production and Engineering Staff are required to be on the set at least one hour prior to technical rehearsal. The Production and Engineering staff will *set up for their area only*. Talent is required to stay out of the production area until the Director is told and set and engineering crews are ready to start shooting. Cameramen, floor audio, and master control staff are the only authorized people allowed to be near or operate equipment, under the supervision of the Chief Engineer and Staff Director. The cameramen for the evening are the only authorized people to operate and man cameras, again under the direction of the Staff Director and Chief Engineer. Lighting, mikes, and other electronic equipment are not to be changed without prior approval of the Director and audio man. Cameramen will assure that their cameras are out of the lights at all times and are not left unmanned or focused on one set for a long period of time. Camera f-stops are not to be changed after they have been set except by the approval of the Director or Chief Engineer.

Since we often have guests in the studio, the entire staff will be required to conduct themselves in an orderly and professional manner. We are with the system to learn and to have fun. Having fun should remain in the realm of good taste.

Questions, comments, or ideas concerning any of the standard operating procedures listed below should be brought to the attention of the Program Director.

Summary of Staff Member Responsibility

Staff Member	Report to
Cameramen, floor manager, lighting, floor audio	Production Supervisor
Audio man, video man, all engineering personnel	Chief Engineer
Directors, staff announcers, department heads, talent, producers and writers	Program Director
Graphics	Graphics Director
Grips, set designers and assemblers	Scene Designer

Appendix B

GUIDELINES FOR ACCESS: A REPORT BY NCTA

August, 1972

The National Cable Television Assn.
918 16th St., NW
Washington, D.C. 20006

The Federal Communications Commission's new rules for cable television pose a new challenge for the cable television industry and for the public. The requirement that cable systems provide non-broadcast designated access channels for public, educational, governmental and other uses begins a significant new aspect of broadband communications services.

This report offers guidelines to cable operators to help them fulfill the FCC's requirements. At the outset it should be clear that it is not possible or even desirable for us to establish detailed recommendations for the operation and development of these new services. The local and unique needs and capabilities of the cable operator and his or her community precludes such advice. This report, then, takes the general approach — identifying what types of operating rules must be set, pointing out issues and problems that should be met, suggesting possibilities and in some cases offering parameters. We are also recommending that systems establish interim operating rules for a specified period of time — six months or a year, for example. At the end of that time a complete review of the rules in light of actual operating experience should be undertaken.

Because there are varying requirements for each of the four categories of designated access channels, we have treated each category separately. (Problems concerning accountability for libel and liability insurance are not treated here; these matters are under separate study.)

If the concept of public access is to work for the cable operator and the community — for all groups and individuals concerned — there must first be a realization by all that these are largely uncharted waters. No one has a great deal of experience upon which to build. The FCC itself has purposely offered only general guidelines.

Success will not come overnight. For the access concept to develop meaningfully, flexibility, persistence and perhaps most important, sincere, honest and realistic efforts at cooperation will be required of all concerned.

David Foster
President
National Cable Television Association

Application of rules

The FCC's access requirements apply to cable television operators as follows:

All systems located wholly or partially within a top-100 television market which began or begin operations on or after March 31, 1972.

Systems already operating in the top-100 markets have until March 31, 1977 to comply. However, such systems providing any of the access services prior to that time must comply with FCC requirements (1) barring operator control over program content (2) assessment of costs and (3) the operating rules. Systems receiving certificates of compliance to add television signals to their operations prior to the March 31, 1977 date must comply with the access requirements, by adding one specific access channel for each broadcast added. Priority for adding channels is public access, first; educational, second; government, third; leased, fourth.

Systems located wholly outside major television markets cannot be required by a local entity to exceed the FCC's access requirements. However, if such a system does provide any access service, it must comply with the requirements.

The FCC is aware that the requirements may impose undue burdens on some cable operators. In such cases, the operator may request a waiver. Cable systems operating in small communities within a major market, to whom the access requirements present a burden, are free to meet their obligations through joint building and related programs with other cable operators in the larger core areas.

Activation of additional access channels

The FCC has required that when use of the designated access channels reaches a specified point, additional channels must be activated to meet the demand. The commission's formula for expansion of access channel capacity is:

Whenever all of the designated access channels (public, educational, local government, or leased) are in use during 80 per cent of the time during any consecutive three-hour period for six consecu-

tive weeks, a system shall have six months in which to make a new channel available for any or all of the above described purposes.

The Public Access Channel

There are six general FCC requirements related to the public access channel. (1) The system must maintain at least one such non-commercial channel. (2) At least one public access channel shall always be free. (3) The system must maintain and have available for public use at least minimal equipment and facilities necessary for the production of programming. (4) The system may not assess production costs for live studio presentations of less than five minutes. (5) The system may not exercise any control over program content (except as mentioned below). (6) Finally, the system must establish operating rules for the channel. A copy of the rules must be filed with the FCC within 90 days of the date of activation of the channel. A copy of the rules must also be available for public inspection.

Operating rules

The system's operating rules must specify the following:

Access is to be first-come, first-served, non-discriminatory.

A prohibition on the presentation of any advertising material designed to promote the sale of commercial products or services (including advertising by or on behalf of candidates for public office).

A prohibition on the presentation of any lottery information (as in the cablecasting rules).

A prohibition on presentation of obscene and indecent matter (as in the cablecasting rules).

Permission of public inspection of a complete record of the names and address of all persons or groups requesting access time. (The record must be retained for two years.)

Those are the specific requirements levied by the FCC. In your published operating rules you will want to deal with other factors, as well.

Production facilities

The FCC requires cable operators to have "at least minimal equipment and facilities necessary for the production of programming." We believe that where economically feasible, provision should be made for a studio in good operating order with at least two cameras, two video tape recorders and the basic attendant studio equipment (lighting, etc.). Beyond this, the amount and type of equipment should be a function of demand. The equipment should be in good operating condition, providing a level of technical quality consistent with the user's offering. Experience thus far indicates it is important to have as high a level of technical quality as possible, in order to attract and maintain the interest of viewers accustomed to good quality. The cable operator may require reasonable technical standards in software submitted by a user. Provision should also be made for the system to provide a qualified television production person to offer technical assistance during reasonable business hours.

Although the commission is requiring that a system have production facilities available, it also expressed the hope that "colleges and universities, high schools and recreation departments, churches, unions and other community groups will have low cost video taping equipment for public use."

Production costs

FCC rules stipulate that you may *not* charge for costs incurred in live studio presentations of less than five minutes. Beyond that, production costs are chargeable to users. With respect to the latter, the FCC has said that production costs and any other fees for use of other public access channels "shall be consistent with the goal of affording the public a low cost means of television access." In compliance with the spirit of that directive, we believe charges to users of production facilities should be limited to actual costs incurred, whenever possible. Basic rates should be set and published in the operating rules.

Formats and allocations

The basic criteria for most decisions in public access should be supply and demand. However, in some cases, cable operators may want to set aside certain days and times to deal with "regular" access offerings and with "one-time" access presentations. There is advantage to the operator, to the user, and to the viewer in permitting at least a limited degree of regularity. But this should not be structured so as to interfere with the one-time user's right to access.

First, the operator will want to establish a reasonable "broadcast day." He may want to require that applications for channel time be made in advance: for example, one or two weeks. (Such a requirement could and should be waived if no demand exists for available time.) Applications should specify, among other things, requested time and date, program length,

purpose of program, persons appearing in program, method of presentation, requirement for studio facilities, etc.

Operating rules might specify limitations on regular and one-time use. For example, a limit on the number of hours of prime time per week per user, a limit on the total number of hours per week, or even a limit on the length of time a user can receive a particular time block. However, limitations on access time should relate to demand. If demand is slight, limitations should be minimal. If demand is heavy, limitations may be more severe. The guiding principle should be to give access to the greatest number of users.

Copyright

Operating rules should specify procedures to insure that copyright clearances have been obtained. To relieve the system from copyright liability, rules could require each user to furnish appropriate documentation that, where copyrighted material is included in the program, clearance has been obtained. The simplest method would be to require written proof of clearance for both music and non-music copyrighted material. In the case of music, the operator might require the user to furnish the title of the music to be used, the name of the composer(s) and the licensing agent for performance rights, and documents authorizing performance. For non-musical material, information should include the name of the author, the copyright owner, and again, documents authorizing use of the material on the program.

Other requirements

There are other issues for which you may want to consider appropriate operating rules. You may want to include a provision dealing with the use of access facilities by minors. You might require that any minor using the access facilities be accompanied by an adult, who would assume legal responsibility for the actions of minors and also for obtaining documents required for use of the access channel.

You might include a provision requiring access users to provide to the cable system any information necessary for the system to meet FCC requirements. Finally, you will probably want to outline procedures to follow if disputes or grievances develop over program content, technical standards, etc.

Promotion of access

Early experience with public access channels has demonstrated that promotion is an extremely important factor in the healthy development of the access concept. The cable operator should involve himself in the promotion and development of the public access channel. Cooperative efforts with community groups to further understanding of the concept of access and effective utilization of access channels could be one approach. In fact, cable operators should consider the formation of broad-based democratically constituted groups to advise and assist in policy development, promotion and funding of public access.

Such a group or groups lend valuable assistance, identifying and assisting potential users, helping to generate community interest in the channel, and developing sources of funding.

The cable operator can contribute to the promotion of the access channels through his ongoing advertising and marketing programs. Notice of the availability of access time and programming schedules can be given exposure on non-broadcast channels and by mailing promotional material with monthly bills.

The Educational Access Channel

The broad requirements of the educational access channel are these: (1) At least one must must be made available free of charge from the commencement of cable service until five years after the completion of the basic trunk line; (2) you may not exercise any control over program content (except as mentioned below); (3) you must establish operating rules for the channel. A copy of these rules must be filed with the FCC within 90 days of the date you first activate the channel. A copy of the rules must also be available for inspection by the public.

Operating rules

The FCC requires that the operating rules specify the following four points:

A prohibition on the presentation of any advertising material designed to promote the sale of commercial products or services (including advertising by or on behalf of candidates for public office).

A prohibition on lottery information (as in the cablecasting rules).

A prohibition on obscene and indecent matter (as in the cablecasting rules).

Permission of public inspection of a complete record of names and address of all persons or groups requesting time. (This record must be maintained for two years.)

Beyond these broad requirements and the four points mentioned above, you are free to develop further and more detailed operating rules. It is in your interest to establish rules and procedures to promote successful development of the educational channel.

Following is a brief discussion of some of the factors you should consider.

Users

The FCC has specified only that the channel is for use by "local educational authorities". The channel is *not* exclusively for the use of any one group — be it the public or educational broadcaster, a school or other educational group. You will want to determine who are the local educational authorities in your community or franchise area and who are the potential users. Prior to activating the channel, you will want to notify potential users of the channel's availability.

You might consider establishment of a local educational access advisory board composed of, for example, a cable representative, an educational broadcaster, a school board member, a student, and an educational or other public official, etc. This board could advise on what groups may utilize the channel and on time allocations when there are competing applications. You should remember, however, you cannot pass on responsibility for enforcing FCC requirements.

Demand may be difficult to guage at first. You may want to adopt a flexible approach for the first year.

Order of access and time allocations

No requirement exists for first-come, first-served access on the educational channel. At the outset you will probably want to establish a "broadcast day". Order of access (and time allocations) will probably be a function of demand. Establishment of priorities may be possible and wise. For example, instructional programming designed as a part of or supplement to school curricula might need to be presented during the school day. Allocation of regular times and days may be advantageous for such programming. If there is a great deal of demand for both regular and one-time or special programs, it may be necessary to establish limits on time allowed per day, week, etc. However, for the near term, that seems unlikely. Access and time decisions should develop according to demand.

Production costs and facilities

You are not required to provide production facilities for the educational channel user. However, in some cases, systems may find they are able to provide such facilities. For example, since systems must have facilities for the public access channel, those same facilities, if not in use, might be used for the educational access channel. The same might be true of regular cablecasting facilities. If systems do provide production facilities, they may charge for use. However, where possible we recommend that charges in this early stage of access development should only be for recovering the actual cost of the service. Rates for such services should be established and published in the operating rules. Since some educational institutions — schools and universities, for example — have production facilities and equipment which are often not fully utilized, cable systems may be able to work out arrangements for use of equipment and facilities.

Copyright

Use of copyrighted material on the educational channel could be a problem for the cable operator. You can and should require appropriate documentation protecting the system from copyright liability. Generally the same information outlined in the public access section should be required.

Funding and promotion of channel

Once the educational channel is activated, a significant factor in its success will be funding to develop software for the channel and promotion of the channel. There are no FCC legal requirements on the operator in these areas. However, operators should be attuned to the need for funds and promotion and should work with educators, community groups, and others in tapping potential sources of funding and in promoting the channel to both users and viewers.

The Government Access Channel

The operation of the government access channel differs markedly from the three other categories of access channels referred to in this report. FCC guidelines on the government access channel are quite general. The commission says the channel is "designed to give maximum latitude for use by local governments". Of the four categories of access channels, the government access channel is the only one in which regulation of access is not precluded.

The FCC's guidance is limited to the following: (1) At least one such access channel must be made available free of charge from the commencement of cable service until five years after the completion of the basic trunk line. (2) The system may exercise no control over program content. (3) Local regulation of the channel is permitted.

Development of this access channel will largely be a concern of the local government. The FCC is not requiring the system to establish operating rules, nor is it requiring the system to make available production facilities. However, we believe that cable operators should assist the local government in setting operating rules; advising, for example, on the establishment of production facilities if utilized.

The Leased Access Channels

After the cable system has met FCC requirements for use of designated non-broadcast channels, it then must make available the remainder of the required band width for leased channel operators. Latitude and flexibility should be key concerns in formulating operating rules for this category of channels.

The FCC has issued the following requirements for leased access service. (1) A system may exercise no control over program content. (2) There are to be no limitations on commercial matter. (3) On at least one leased channel, priority shall be given to part-time users. (4) A broadcast channel blacked-out because of exclusivity rules may be used for leased channel purposes. (5) Unused portions of the specially designated channels (public access, educational, etc.) may be used for leased service, but such service is subject to replacement when a demand exists for the designated use of that channel. (6) The system must establish operating rules. A copy of these rules must be filed with the FCC within 90 days of the date of activation of the channel. The rules must also be available for public inspection.

Operating Rules

The system in its operating rules must specify the following:

Access shall be first-come, first-served, non-discriminatory.

A prohibition on the presentation of lottery information (as in the cablecasting rules).

A prohibition on obscene or indecent matter (as in the cablecasting rules).

Sponsorship identification (as in the cablecasting rules).

Appropriate rate schedule.

Permission of public inspection of a complete record of the names and addresses of all persons or groups requesting access time. (This record must be maintained for two years.)

Access and allocations

Access to the leased channels must be first-come, first-served. Deadlines for applying for a specified access time will probably not be necessary, at least for the near term. A possible exception to this might be in the case of leased channels for part time users, and also in the case of leasing time on blacked-out broadcast channels. Here deadlines and limitations on the amount of time, time spots, etc. might be appropriate, if demand requires it. Only experience and time will tell.

On all other leased channels there appears to be no need for limitations on time allocations. Supply and demand should be the governing factor here.

Rates

Cable systems must establish rates for channel leasing and include them in the operating rules. Again, flexibility is important. But it will be up to the individual operator to assess his particular situation and develop the appropriate rate schedule. Rates should vary, for example, for full use and part-time use; also, for prime time and non-prime time hours. Different rates might be established for commercial and non-commercial users. Rates should also vary for leased operations on designated access channels and blacked-out broadcast channels.

Production facilities

Operators are not required to provide production facilities for users of leased channel operations. But, as in the case of the educational channel, operators may find they are able to provide these facilities. If this is done, appropriate rates should be established.

Advertising

There are no restrictions on advertising on leased channels (unlike the local origination channel). As the FCC notes, some channels may be used entirely for advertising. The commission has also indicated that it will monitor developments in this area, particularly in regard to issues such as false and misleading advertising. Because cable operators are not allowed any control over program content on designated access channels, generally they cannot control advertising. However, systems can adopt an advertising and merchandising code of ethics (the NCTA cablecasting code of ethics, for example) and acquaint users with the code and request that they conform to it.

Copyright

Generally the same problems that exist with respect to copyright liability for other access channels apply to the leased channels. The steps outlined in previous sections on copyright apply here.

Appendix C

ORGANIZATIONS WORKING IN PUBLIC ACCESS

This list of organizations working in public access is not intended to be complete or exhaustive. Many small local organizations have extensive interest, experience, and knowledge in cable generally and public access in particular; however, many are too small to handle large numbers of inquiries. They are too numerous to list here, but can be found in the publications listed under "References" below.

Alternate Media Center
New York University
School of the Arts
144 Bleecker Street
New York, New York 10012

> The largest and best-funded public access resource in the United States. It has developed projects all around the country.

Bedford-Stuyvesant Restoration Corporation
1368 Fulton Street
Brooklyn, New York 11216

> Conducted detailed study and plan for a community CATV system in an area with 500,000 people.

Black Efforts for Soul in Television
1014 North Carolina Avenue, S.E.
Washington, D.C. 20036

> Primarily broadcast oriented, BEST aids community groups with programming complaints, license challenges, technical assistance, employment, and so on.

Cable Television Information Center
2100 M Street, N.W.
Washington, D.C. 20037

> A Ford Foundation supported organization geared toward providing municipalities with a broad range of cable information.

Citizens Communication Center
1812 N Street, N.W.
Washington, D.C. 20036

> Provides legal assistance to the public before the FCC and the courts on communications issues.

Fide's House/Community Video Center
1554 8th Street, N.W.
Washington, D.C. 20009

> Neighborhood video information, production, and training center.

National Cable Television Association
918 16th Street, N.W.
Washington, D.C. 20006

> The NCTA, an industry group, issues reports to its membership on cable issues. The NCTA Bulletin, published every other week, is available to nonmembers for $24 a year. It can provide resource and technical information, and has issued a useful set of guidelines for public access (reproduced above as Appendix B).

Open Channel
49 East 68th Street
New York, New York 10003

> An experienced production and information service for public access.

People's Video Theater
New York, New York

> One of the first video groups geared to community service, and one of the few actively pursuing the use of video as a neighborhood information tool.

Port Washington Public Library
Port Washington, New York 11050

> A functioning access center, not as yet involved in cable but with expertise in the "process" use of video.

Public Advocates
433 Turk Street
San Francisco, California 94102

> A public interest law firm with experience in cable television.

Reading Community Video Workshop
Berks TV Cable Company
P.O. Box 107
Reading, Pennsylvania 19603

> An example of cooperation between a community access center and a cable system.

Stern Community Law Firm
2005 L Street, N.W.
Washington, D.C. 20036

> A public interest law firm involved in broadcast and cable television.

United Church of Christ
Office of Communication
289 Park Avenue South
New York, New York 10010

> A leading proponent of public interest in broadcast and cable television affairs, with several useful publications.

Urban Communications Group
1725 De Sales Street, N.W.
Washington, D.C. 20036

> A black consulting firm with extensive cable television experience, particularly interested in minority ownership of media.

Watts Communications Bureau/Mafundi Institute
1827 E. 103rd Street
Los Angeles, California 90002

Offers video training for third-world people. The first group to propose and plan a community-owned cable system (the franchise is still under negotiation).

REFERENCES

The following list comprises what the author regards as the best of the literature on public access. Each item will direct the reader to additional sources. For a more comprehensive bibliography of cable writings, see "For More Information" in Walter S. Baer, *Cable Television: A Handbook for Decisionmaking,* The Rand Corporation, R-1133-NSF, February 1973.

Broadcasting
1735 De Sales Street, N.W.
Washington, D.C. 20036

(weekly, $20/yr)

> Covers both broadcasting and cable, with emphasis on FCC matters. An annual CATV Sourcebook is also available for $8.50.

Cable Television in the Cities: Community Control, Public Access, and Minority Ownership, Charles Tate (ed.), The Urban Institute, 1971, 184 pp. ($3.95).

> An excellent manual with an extremely detailed resource and reference directory. Available from The Urban Institute, 2100 M Street, N.W., Washington, D.C. 20037.

Cable Television: Opportunities and Problems in Local Program Origination, by N. E. Feldman, The Rand Corporation, R-570-FF, September 1970, 31 pp. ($2.00).

> A useful report on a number of local origination experiments, including personnel requirements, costs, and administrative structure.

Cablecasting
140 Main Street
Ridgefield, Connecticut 06877

(bimonthly, $8/yr)

CATV
1900 W. Yale
Englewood, Colorado 80110

(weekly, $33/yr)

> The annual **CATV Systems Directory** ($8.95) and **CATV Directory of Equipment and Services** ($6.95) are free with a **CATV** subscription.

Challenge for Change Newsletter
National Film Board of Canada
P.O. Box 6100
Montreal 101, Quebec

The newsletter of the pioneers in community television. Easy to read, filled with graphics and details of forceful video and cable projects. Three or four issues yearly. Available free.

Cinema's Catalyst: Film, Videotape and Social Change, by Sandra Gwynne, Extension Service, Memorial U., St. Johns, Newfoundland.

A reporter's view of a conference on community uses of media, including a list of participants.

Radical Software
440 Park Avenue South
Suite 1304
New York, New York 10016

The journal of the Alternate Video movement. The back issues are particularly useful as an introduction to video and cable and as a source guide to video groups and organizations around the world.

TV Communications
1900 W. Yale
Englewood, Colorado 80110

(monthly, $10/yr)

Contains more in-depth articles on cable than do other journals listed.

Part II

Applications for Municipal Services
Robert K. Yin

I. INTRODUCTION

Cable television's growth in the top urban markets may depend heavily on the new services it delivers. Unlike rural systems, which have prospered by bringing television signals to viewers inadequately served by broadcast stations, urban cable systems will have to compete with the good over-the-air reception prevailing in most cities. Logic thus decrees that new programs or services will be needed to attract a large subscriber base.

These new programs or services can be extensions of existing commercial programming, such as the cablecasting of movies and sports events unavailable over-the-air. Alternatively and in addition, they can be entirely new public services directed toward community and municipal needs. Most recent studies of the future of urban cable have included extensive discussions of public service applications; the "television of abundance," as cable television has been called, can improve the quality of urban life only if the abundance offers diversity in entertainment *and* ample new public services.[1]

Meanwhile, municipal officials face a specific challenge in the near future. The FCC rules require all new cable systems in the 100 largest television markets to dedicate a single channel for municipal use, free of charge for five years.[2] The requirement provides an immediate opportunity to test new municipal programs, and possibly to create a demand for the allocation of additional municipal channels at a later date. (The current FCC rules also permit leasing of additional channels for municipal use.) At the same time, if the municipal channel is not used effectively or is grossly underused, the FCC may decide to eliminate it and allocate the channel space for other purposes. In short, the time has come to examine closely the myriad ideas and the few operating experiences with cable television and municipal services, and to determine what services municipalities can provide by cable.

There are many possibilities. Municipal officials will need to examine the issues and problems at three levels:

---

[1] See, for instance, Committee on Telecommunications, *Communications Technology for Urban Improvement,* National Academy of Engineering, Washington, D.C., 1971; Sloan Commission on Cable Communications, *On the Cable: The Television of Abundance,* McGraw-Hill, New York, 1971; and William F. Mason, et al., *Urban Cable Systems,* The MITRE Corporation, Washington, D.C., 1972.

[2] Federal Communications Commission, *Cable Television Report and Order,* 37 Fed. Reg. 3252, February 12, 1972. The FCC regulations are outlined in Baer, op. cit., and discussed more fully in Rivkin, op. cit.

- What applications are possible for different types of cable systems;
- The general feasibility and usefulness of each application; and
- The specific feasibility and usefulness of each application for a particular municipality.

This part of the report deals with only the first two levels. Chapter II deals with the major choices involved in using the single municipal channel. Chapter III discusses applications on advanced multichannel systems. Chapter IV reviews specific applications for several municipal services: public safety, criminal justice administration, environmental monitoring and control (including traffic and transportation), health care, social services, and regional communication. Since the purpose is to discuss as many applications as possible, their individual treatment is necessarily brief; numerous footnotes direct the reader to sources of further information. One major service area, education, has been purposely omitted here because it has so many applications for cable that they are the subject of two separate parts in this book.[3] Finally, no attempt has been made to cover issues at the third level, since only municipal officials themselves can account for local variations in service delivery and service needs.

The following are some of the major conclusions stemming from the review:

- Most applications will require new resources from the municipality, in the form of equipment, operating expenses, new programming, and appropriately skilled staff.
- Because most municipal applications have not been field-tested, little is known about their actual feasibility and cost.
- Most applications that involve a significant change in service will require a two-way communication capability, a feature that urban cable systems are not likely to have in the immediate future.

The last point leads to an important observation: the experience derived from operating a single municipal channel may not be very useful in assessing the need for more advanced cable systems. This is because a single channel is confined to only a few types of one-way programs. Most will involve public education and public relations programs aimed at a widely dispersed audience, and inservice training programs aimed at such special users as policemen or doctors. Only when interactive multichannel capabilities are available can such applications as fire and burglar alarm systems, street surveillance systems, and individualized health and welfare services become possible (and even these applications must be adequately tested and demonstrated). Single-channel and advanced systems also will require considerably different media management. The municipality is likely to operate the single channel as a *broadcast* facility; it is likely to operate the multichannel, two-way system as a *communications* facility.

The main sources of information for this report's discussion are the numerous working papers and research studies that have accumulated over the last few years. Even though these materials are rather thin, they have been cited wherever possible

---

[3] See Part III and Part IV.

(especially in Sec. IV, which describes applications individually), so that the reader can refer to them for further information. The discussion also draws on the working experiences of one municipal television station, WNYC-TV (New York City), although it operates on a UHF frequency and the parallels between its experiences and the likely cable experiences may be limited. In particular, WNYC-TV has not developed good information about its audience characteristics. On the other hand, it has been in operation for just over ten years, and is the only municipal television station in the United States. (The Appendix describes the station's activities.)

II. OPERATING A SINGLE MUNICIPAL CHANNEL

The municipal channel will allow city governments to provide improved services to citizens. The types of services are discussed in Section IV, but regardless of the specific types, the operation of the channel raises some important and difficult questions. The following discussion reviews these questions and offers guidance on possible answers; the reader should also refer to the discussion of WNYC-TV in the Appendix.

CONTROL AND ACCESS

A basic issue is control over the use of the channel. In most cases, control will rest with the executive branch of government, which means that office-holders could use the channel not only for service delivery, but for politically motivated programs. The limits of such political uses have not been tested to date, nor is it clear whether the executive and legislative branches of local government can or should share in the operation of the municipal channel. In New York City, the executive branch operates WNYC-TV and has avoided overt political programs. At the same time, the municipal channels on New York's two cable systems will be operated under the Board of Estimate, a part of the legislative branch.

FCC regulations are relatively unrestrictive about the use of the municipal channel. In some cases, as in Massachusetts, state governments may eventually play a more dominant regulatory role. Even with outside regulations, however, municipal officials must decide for themselves the most appropriate arrangement that: (a) conforms to local legislation and regulations, and (b) produces an equitable means of distinguishing politics from service delivery, with as much emphasis on service delivery as possible. Even among service delivery alternatives, whoever controls the municipal channel will have to exercise sound judgment. And if, for instance, officials decide to emphasize health programs, they may want to make their selection criteria as explicit as possible.

SITES

A second important question should be settled before the cable system is installed: the extent to which the municipal channel should merely be part of the regular home delivery system, or whether and how the cable system should also cover existing municipal facilities. Home delivery would be like an extension of conventional TV, with programs transmitted to residents in their own homes. If programs are planned for inservice training of policemen, health professionals, or teachers, however, connections should be made to precinct houses, hospitals, and schools. This means creating the communications link and designating reception areas within these facilities. Most current urban cable systems routinely include municipal facilities. Officials should make sure that all the relevant sites, including those for the near-term future, are covered.

An alternative to home delivery is cable connection to a set of *neighborhood centers*, which have several virtues.[1] First, many people who can benefit most from the services of the municipal channel may not subscribe to the cable system. They might be willing and able to use a common public facility, however, which might be located as close as the lobby of their apartment building or in a nearby storefront office. Second, the neighborhood centers could combine telecommunications with other services; for instance, information broadcast over the municipal channel could be combined with personal attention from a paraprofessional, with access to forms that need to be filled out, or with access to the resident's official records so that further action can be taken. None of these auxiliary services are possible in the home delivery system. Third, the neighborhood centers could be the site for the use of more advanced technology, too expensive for home use. For instance, neighborhood centers could originate their own signals, and thereby have a two-way audio-video link with other public facilities. A two-way capability would enable the development of a host of different services (see Sec. III).

No existing cable system fully tests the neighborhood center idea. Municipal officials should explore it before the cable system is installed by considering which new or existing facilities might be connected to the cable as neighborhood centers. Because new sites may be too costly, officials may want to consider using branch libraries as neighborhood centers.[2] Branch libraries have the virtue of being dispersed throughout a city, and most people know where they are. In addition, since libraries in many cities are becoming underused, municipal officials may already be looking for new uses for them. Naturally, it will take careful study of local conditions to determine the compatibility of library services with services that might be provided on the cable system.

PROGRAMS

Many types of programs are suitable for single-channel use, and municipal officials will have to decide which programs and how much programming time to

---

[1] For one discussion of neighborhood centers, see Alan R. Siegel and Calvin W. Hiibner, "Special Services to Neighborhood and Home: A Community Telecommunication Demonstration Concept," paper presented at the International Telemetering Conference, October 1972.

[2] Port Washington, New York, has done this with great success. Although the town has no cable system as yet, the library houses a sizable collection of videotapes produced by local citizens. Enthusiastic audiences of townspeople assemble at the library to view these tapes. See Part I.

use. As a general guide, most municipalities may find it advisable to start off with only a few programs and make increases gradually.

Section IV discusses some specific types of programs. For policymaking purposes, it may be useful here to divide the potential programs into three categories: local government activities, inservice training, and community information programming. A municipality can begin with programs in all three categories, but each has slightly different requirements.

Local government activities would include broadcasts of council meetings, special dedication ceremonies, and other public events. Activities could be emphasized that are in theory public and have substantial local interest, but are poorly attended because of travel or other restrictions. Special hearings by local commissions might be an example. The activities could be covered live, or could be videotaped for later broadcast. Coverage of such events is relatively easy, and could become an important part of the municipal channel's programming day. The programs could be made even more interesting by having question and answer sessions (via telephone and television) with an expert present in the studio after the broadcast.

Inservice training programs would be directed toward municipal officials themselves. They might include programs bearing important information for local administrators, disseminated from a central point to all field installations, again with telephone feedback for subsequent questions. Inservice training programs require more elaborate planning than does mere coverage of local government activities. The nature of the training program and its contents must be prepared ahead of time; the officials to whom the training is directed must be scheduled for availability, and training sites have to be made available. Whenever possible, such programs should be broadcast at hours that do not conflict with the best viewing hours for programs directed to the general public.

Community information programming would be aimed at assisting citizens and their use of municipal services. The most attractive programs would use visual displays, and not rely on words alone. Programs that might attract special interest would be demonstrations of the bathing, feeding, and care of young infants; do-it-yourself household repairs; home safety precautions (with demonstration of fire and other hazards), and lessons for first aid in the home. Programs dispensing employment or other information might be less interesting unless augmented by unusually creative video highlights of different occupations in action.

Much community information programming may be purchased from media and professional groups.[3] For specific service functions, such as housing, health, or welfare, the officials in the operating agencies of the local government may have greater knowledge about suitable programming than will officials, say, in the mayor's office, and should therefore be included in the decision process.

COSTS

Unless the municipal government has its own video production studio, it will

---

[3] Since these groups are scattered and not easy to locate, municipal officials should keep in touch with the Cable Television Information Center, 2100 M Street, N.W., Washington, D.C., 20037. For a compendium of other sources of information about cable, see "For More Information" in Baer, op. cit., p. 221.

have to use the cable operator's facilities or those of another institution (such as a local community college). Building a new studio probably will be too expensive; the minimal facility needed for a six-hour operating day could require about three persons, which would mean an annual personnel cost alone of over $30,000. The municipality probably will also have to augment the media staffs in its various operating agencies. (How many totally new personnel are required will depend, of course, on the ability of existing personnel to perform the new cable-related functions.) The operating agencies might have to augment their staff at a minimum by two to three persons each, and thus perhaps spend an additional $30,000 a year each for personnel costs and program materials. Assuming that four agencies (e.g., police, fire, health, and social services) produce programs in a city-run studio, use of a municipal cable channel could thus cost the municipal government about $150,000 more in operating costs alone.

In some ways, such an estimate may be conservative, since New York City's WNYC-TV has a budget of about $700,000 per year, supporting about 40 staff members, and with a studio of its own. On the other hand, municipal governments might manage to use the cable operator's studio, and might also choose to emphasize live or taped coverage of public events, such as city council meetings, public hearings for zoning changes, courtroom cases, and school board meetings. Coverage of these events would provide new information not generally accessible to the mass public. Such an emphasis could reduce the operating costs below $100,000 a year, but the programs would only enhance the visibility of public events, and would not affect the delivery of municipal services directly.

One issue often raised is whether the municipal channel could actually lead to any reduction in municipal expenditures. It could, if the channel replaced existing and more expensive services, but this is not likely to occur in the near future. First, the functions it might replace in most agencies are public and community relations activities, which do not usually entail high costs. Second, in most cities, the use of television in public services is a new function, requiring differently trained personnel and new equipment. Given the way bureaucracies work, it will be easier to purchase new equipment and hire new personnel, rather than attempt retooling and retraining. Third, there is no evidence yet that television can enhance service effectiveness; and even if it did, the improvement is likely to result from satisfying previously unmet service demands—in which case services would expand, not contract.

III. CABLE SYSTEM CHARACTERISTICS RELEVANT TO MUNICIPAL APPLICATIONS

The full range of municipal applications would require cable systems more advanced than a single municipal channel. Use of the single channel as a one-way audiovisual communication link merely assumes the existence of: (1) a *transmitter* at the headend of the cable system, (2) a *transmission channel* linking the headend with individual homes, and (3) a *receiver* at each home. And, in general, cable systems, like any other communication systems, can be specified according to these three elements of *transmitter, transmission link,* and *receiver* (Fig. 1). The major feature of more advanced cable systems is the addition of a two-way audiovisual transmission capability.

These three elements will constrain applications in different ways and to different degrees. Applications like fire and burglar alarm systems require specialized transmitters, but only a minimum two-way transmission channel. Other applications, such as video monitoring systems, may use existing television cameras and monitors as transmitters and receivers, but require numerous two-way channels to maintain constant surveillance of all cameras. Still other applications, such as a

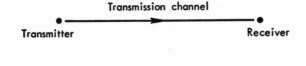

Fig. 1—Basic service system (one-way communication)

public information broadcast program, can use a combination of cable television and telephone links (see Fig. 2).

Since the cable systems currently being built or being contemplated for the near future have known characteristics, it is worth reviewing the municipal applications in terms of these cable systems. In particular, some systems will automatically rule out whole classes of applications, and municipal officials should be aware of these consequences. For further reference, the reader may want to turn to several existing studies of municipal cable systems in specific cities, including the question of municipal service applications. The sites and sponsors of some of these studies have been

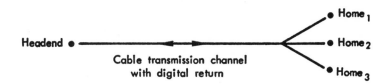

A. Burglar and fire alarm system

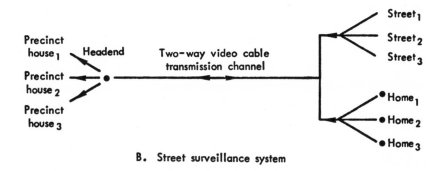

B. Street surveillance system

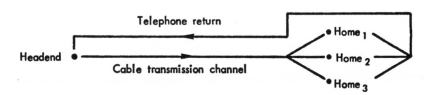

C. Public information program with questions and answers

Fig. 2—Advanced service systems

Dayton, Ohio (The Rand Corporation), Washington, D.C. (The MITRE Corporation), Detroit (Detroit city government), Jacksonville, Florida (Cable Television Information Center), and Cincinnati, Ohio (Wilder Foundation, Cincinnati).

The following discussion begins with a conventional, one-way cable system that has been and will likely continue to be installed in the near future in urban areas. The discussion then covers four possible additional functions to the basic system: private channels (achieved with special receivers), digital two-way communications, audiovisual two-way communications, and special transmitting devices.[1] Private channels and special transmitters make additional demands on the receiving and transmitting portion of the cable system; digital and audiovisual two-way systems increase demands on the transmission capacity of the cable system as well as on the receiving and transmitting portions. The basic cable system can be augmented by a combination of these additional functions; and as we shall see, different municipal applications call for different combinations of additional functions.

CONVENTIONAL, ONE-WAY CABLE SYSTEM

Today's basic cable system carries up to 26 channels per cable, involves one-way communication (though the FCC rules require that it have the capacity for eventual two-way communication of some sort), and makes no discrimination among the audience (any person connected to the cable system can receive any program being cablecast). The local transmitters are the television cameras and other origination equipment in the studio, and the receiver is the television set in the home or office. The cable systems to be built in the next few years will set aside one channel for municipal service use, and additional channels will be available on a leased basis. For this basic system, only public educational and inservice training applications will probably be appropriate; these would include cablecasting of public events such as public hearings, public safety education, public legal information, public health education, social service information, and the like.

ADDITION 1: PRIVATE CHANNELS

With different types of receivers (e.g., a converter attached to the television set), the audience can be differentiated, with only some homes able to receive certain channels. In effect, this arrangement enables the cable system to have private channels, with programs available only to a restricted audience and under certain conditions. Pay TV, for instance, would require this type of arrangement. For municipal services, private channels might be needed for instructional programs aimed at lawyers, policemen, or doctors, or for any specialized inservice training. Where audience limitation is unnecessary, the basic system alone will suffice. In New York City, for instance, police training programs are broadcast even though WNYC-TV is a UHF station and virtually anyone can view the programs.

[1] Full descriptions of the system configurations can be found in Baer et al., op. cit.

The most extreme form of a private channel is an internal closed-circuit TV system, operating within a single set of facilities. Such internal systems would be used for prison surveillance or for the supervision of day care and nursing home facilities. These systems could be built independently of the cable system, or given a two-way communication link with other terminals.

ADDITION 2: DIGITAL FEEDBACK

Digital feedback would provide a minimal level of two-way service, and would open the possibility for any municipal applications requiring low-capacity data feedback. For instance, it could be used to poll homes in an electronic referendum, or to determine what homes are watching which programs. Other applications embrace an array of environmental monitoring systems, including fire and buglar alarm systems, traffic monitoring, pollution monitoring, and utility meter reading.

ADDITION 3: AUDIOVISUAL FEEDBACK

Full two-way communication would have audiovisual signals returning upstream on the cable. Such a system would enable applications requiring any combination of audiovisual downstream and audiovisual or digital upstream flow. For instance, a video surveillance network of the streets could involve digital signals from the precinct house to the cameras on the street (to direct camera movement and focus), and audiovisual signals returning from the street cameras to the precinct house. Interpersonal communication between municipal offices would also require a two-way audiovisual capability.

ADDITION 4: SPECIAL TRANSMITTING DEVICES

An additional feature not generally considered part of the cable system is nevertheless important from the standpoint of municipal applications. Several applications, such as fire and burglar alarm systems and other environmental monitoring systems, require specialized sensors or detectors as transmitting devices. Many of these sensors have not yet been developed, or are expensive; and even those that are available and cheap must be used in such great quantity (e.g., burglar detectors might have to be located at every point of entry in a house) that installation and wiring could be a problem.

In considering the full range of applications, then, it is necessary to keep in mind the potential requirements for special transmitting devices. In other words, even if a cable system had a minimal two-way digital feedback capacity, a fire or burglar alarm system could still not be implemented unless adequate transmitting devices were available and could be installed easily.

SUMMARY OF APPLICATIONS ACCORDING TO CABLE SYSTEM CHARACTERISTICS

Table 1 summarizes the range of municipal applications according to the cable system features they will require, and therefore can be used in selecting both systems and applications. The applications are grouped into service areas, and are in the order in which Sec. IV discusses them.

Table 1 illustrates the limited range of applications that will be possible under the basic one-way, one-channel cable system. Single-channel programs will tend to concentrate on educating the citizenry, as in public safety, public health, and public legal information programs, and on covering such public events as court hearings and school bond meetings. Other programs can present inservice training, to the extent that special converters are not needed to limit the audience.

The operations of the single municipal channel, then, may resemble the operation of local television stations, with central studio facilities and a small technical and production staff. In addition, there will need to be media counterparts in the major operating agencies of the municipal government, so that the production of public education programs, for example, will be a joint effort between an agency and the central studio facility. WNYC-TV operates in much this way, broadcasting a combination of public education programs, inservice training programs, and reruns from the Public Broadcasting Corporation. Clearly, the operation will differ vastly from the multichannel, two-way operation required by more sophisticated services, in which the central facility will be more a communication center than a broadcasting studio.

As a final note, it is important that officials know how many people watch the municipal channel, and their background characteristics and interests. This information can be collected periodically through telephone surveys and the like, and be used to adjust programming according to audience interest and response.

FIELD TESTING OF APPLICATIONS

Table 1 indirectly provides some guidance for planning field tests and demonstrations. The planning for sample field tests in specific locales has already been described briefly in two earlier studies, but with no clear indication of what was being tested.[2]

Broadly speaking, it seems more important to design field tests according to variations in cable system characteristics (basic system, special reception devices, two-way video, etc.) than according to public service characteristics (public safety, health care, etc.). We need to know, for instance, the problems in operating any large-scale, two-way, audiovisual system, whether used for regional communication, personalized health and social services, or interagency communications. Accordingly, one would choose applications from Table 1 so as to field test a variety of cable system characteristics.

---

[2] Committee on Telecommunications, op. cit., and Mason et al., op. cit. On the other hand, a proposed plan for a nationwide test of cable services provides an excellent rationale and discussion of various communication configurations; see Malarkey, Taylor, and Associates, "Pilot Projects for the Broadband Communications Distribution System," Washington, D.C., November 1971.

Table 1

MUNICIPAL APPLICATIONS AND CABLE SYSTEM CHARACTERISTICS

Municipal Application	Basic One-Way System, Single Channel	Special Converters to Restrict Audience	Additional Functions — Two-Way Transmission: Digital	Additional Functions — Two-Way Transmission: Audiovisual	Additional Functions — Special Transmitting Devices
Public safety					
Video surveillance of streets	X	X	X	X	
Fire and burglar alarm systems	X		X		X
Interagency command and control center	X	X	X	X	
Inservice training for law enforcement officials	X	X		X	
Intra-agency daily communication	X	X	X		
Public safety education	X	X			
Criminal Justice Administration					
Courts processing	X	X	X	X	
Security in correctional facilities	X	X	X	X	
Training programs for prisoners	X	X			
Legal services training for lawyers	X	X			
Public legal information	X				
Environmental Monitoring and Control					
Utilities	X		X		X
Transportation	X		X		X
Pollution	X		X		X
Sanitation	X		X		X
Health Care					
Telemedicine	X	X	X	X	
Intra-agency daily communication	X	X	X	X	
Inservice training for health officials	X	X			
Continuing medical education	X	X			
Public health education	X				
Social Services					
Extending person-to-person services	X	X	X	X	
Intra-agency daily communication	X	X	X	X	
Inservice training for social service officials	X				
Community centers	X	X	X	X	
Social service information	X				
Regional Communication					
Interagency communication	X	X	X	X	
Facsimile mail	X	X	X	X	
General Household (polling of homes)	X		X		

NOTE: X = characteristic is required by application.

IV. A REVIEW OF MUNICIPAL APPLICATIONS

This section briefly discusses the range of municipal applications for cable TV. The major purpose is to give the reader an idea of the breadth of applications, rather than to discuss individual applications in great detail. Research papers giving fuller descriptions are referenced throughout, but the reader should be forewarned that the literature on this topic is skimpy and few applications are well understood. The discussion focuses on each application and its required cable system characteristics, and suggests where other communication media, such as telephone or radio, could substitute for cable.

PUBLIC SAFETY

Four representative applications of cable to public safety are discussed below: video surveillance systems, fire and burglar alarm systems, agency communications, and public safety education. Surveillance requires a multichannel capacity and two-way capabilities. Fire and burglar alarm systems occupy only a small portion of channel space, but nevertheless require digital feedback and a possibly substantial investment in the development and installation of transmitting (sensor) devices. The third and fourth applications—agency communications (especially for inservice training) and public safety education—are the most likely candidates for immediate use on the single municipal channel.

Video Surveillance Systems

The rise of crime in the streets has dominated recent concerns for public safety. As a result, perhaps no other cable application has created as much interest as street surveillance by a network of television cameras.[1] In this idealized network, television cameras would be placed at every other street corner for a given area. The television signal would be relayed via cable to a police precinct house, where patrolmen would monitor the screens. Similar systems could operate in public housing facilities, in the public areas of any housing development (lobbies, elevators, corri-

---

[1] A general street surveillance system is described in Committee on Telecommunications, op. cit., pp. 116, 123.

dors, play areas), and in commercial areas like shopping centers. The effectiveness of such systems in preventing and reducing crime has not yet been established, however, and all of them share several important shortcomings.

First, surveillance raises the inflammatory issue of the invasion of privacy.[2] Some would argue that increased safety is worth the drawbacks, because privacy without safety is a nightmare.[3] But local residents have already objected to a field experiment that used eight television cameras for monitoring; after some heated political controversy, the system was eventually dismantled.[4] Americans are exceedingly sensitive on this point. As Baer has noted, "Even a recent proposal to place a government-operated radio receiver in every home for emergency warning was denounced on all sides as a possible first step toward mass surveillance."[5]

The privacy issue is obvious: street and building surveillance may reduce crime, but it also enables the police or others to spy on any person's activities indiscriminately. The situation is uncomfortably reminiscent of George Orwell's spectre of the future in 1984:

> ... The telescreen received and transmitted simultaneously. Any sound that Winston made, above the level of a very low whisper, would be picked up by it; moreover, so long as he remained within the field of vision which the metal plaque commanded, he could be seen as well as heard. There was of course no way of knowing whether you were being watched at any given moment. How often, or on what system the Thought Police plugged in on any individual wire was guesswork. It was even conceivable that they watched everybody at the same time. But at any rate they could plug in your wire whenever they wanted to. You had to live—did live, from habit that became instinct—on the assumption that every sound you made was overheard, and, except in darkness, every movement scrutinized.[6]

(It is ominous, perhaps, that the fictional system of the Thought Police is described in terms of wires and plugs, the very hardware of cable television.)

Before surveillance systems become acceptable, effective control over monitoring agents will be necessary, to assure that they do not watch specific people or places without sufficient and valid reason. It is difficult, however, to conceive an adequate procedure in which monitoring activities are openly known and controllable. One recourse might be to allow citizens themselves to do the monitoring. Citizen monitoring of the streets through cable TV would merely mechanize an existing and normal social function, whereby residents already keep their eyes on the street and informally "monitor" their neighborhoods.[7] It is interesting that there is no evidence of this recourse having been considered.

---

[2] For a discussion of the privacy issue in relation to municipal and public agencies, see Alan F. Westin, "A Preliminary Examination of the Legal and Civil Liberties Aspects of Employing Surveillance Technology to Increase Safety in Public Housing," appendix to William Fairley, Michael Liechenstein, and A. F. Westin, *Improving Public Safety in Urban Apartment Dwellings,* The Rand Corporation, R-655-NYC, June 1971. For a general discussion of the problem of privacy and cable television, see Jerrold Oppenheim, "The Coaxial Wiretap: Privacy and the Cable," *Yale Review of Law and Social Action,* Spring 1972, Vol. 2, pp. 282-288.

[3] Margaret Mead, *Redbook Magazine,* April 1965, as cited in Westin, op. cit.

[4] Gilbert Cranberg, "Cable Television and Public Safety," report prepared for the Sloan Commission on Cable Communications, May 1971, pp. 5-8.

[5] Baer, op. cit., pp. 173-174.

[6] George Orwell, *Nineteen Eighty-four,* Harcourt, Brace and World, New York, 1949, p. 2.

[7] See, for example, the excellent discussion of sidewalk safety by Jane Jacobs in *The Death and Life of Great American Cities,* Random House, New York, 1961, Chapter 2.

Second, the surveillance system has not yet overcome several major technical difficulties. The network would require cameras that either do not yet exist or that are costly (possibly over $15,000 each, including the cost of making them burglar-proof and vandal-proof). Few cameras can operate well under all lighting conditions (from daylight to dark, including those dawn and dusk hours when the sun is low in the sky). Moreover, most cameras have a fixed focus and monitoring, which restrict the range of coverage at the street corner.

Even with a zoom lens and automatic 360-degree rotation, it is unclear how effectively the street can be watched. According to one law enforcement official, the camera's lack of mobility would be a major shortcoming because a suspect could easily flee out of camera range.[8] In addition, before cameras are placed at street corners, analyses should be performed to show where crimes "in the streets" usually occur. If most of them actually occur in alleys, for instance, street-corner cameras will cost far more than they are worth. Finally, such cameras could displace crimes into areas out of camera range, although this could actually be used to advantage. An alternative strategy might be to place cameras in the difficult-to-watch areas of the street, causing any displacement to be in the direction of the more visible areas, which can be monitored more easily by patrol cars or citizens.

Third, the system can require a large number of television channels and two-way capabilities. In a full surveillance system, covering streets on a 24-hour basis, every street-corner camera would require its own television channel, and the resulting channel requirements could exceed the capacity of the whole cable TV system. Sampling only a few cameras at a time would reduce the need for channel space, but would reduce street surveillance proportionately. One possibility is to operate the system only in emergencies. For example, six cameras in a small area might be activated by a street alarm or telephone call from a victim of a street crime. This arrangement, however, can still require many channels, as provision would have to be made for false alarms occurring simultaneously with real alarms, a situation an intelligent criminal would intentionally create.

Fourth, a video surveillance system would impel changes at the precinct house, where several shifts of policemen would be required to monitor the television screens. There have been few tests of policemen's motivation and effectiveness in monitoring, and these would be vital to the system's success.

In conclusion, video surveillance systems not only raise valid social issues such as the invasion of privacy, but would, in any event, have to overcome several technical difficulties before they could be used on a large scale. Adequate demonstration of this application will be needed before municipalities can seriously consider its use.

Fire and Burglar Alarm Systems

Home fire and burglar alarm systems could be a second major application of cable TV for public safety. Both use sets of detectors for heat or smoke (in the case of fire) or for physical intrusion (in the case of burglary). The detectors may be placed throughout the home or any other structure, including stores and offices. When the detectors are set off, a signal is relayed via the cable to the fire or police department

---

[8] Cranberg, op. cit., p. 11.

or a central dispatching office; since each home has its own signal code, the responding units can go immediately to the right place.

In contrast to the street surveillance systems, fire and burglar alarm systems impose less demand on the cable TV facilities. The alarm signals require far less information than do televised pictures of the street, and thus require very little channel space. In fact, the alarm systems may operate very well merely with telephone lines, in which case cable TV will be beneficial only if adequate cable lines can be leased more easily and cheaply than telephonic or other links.[9] Ultimately, cable-connected alarm systems could replace fire alarm street boxes and other alarm devices outside the home as well. The street fire alarm box could become a general alarm box on which a citizen could call any emergency service. The availability of cable TV would mean that the citizen's call could eventually be conveyed with audio and video signals.

The major difficulty with these systems, however, lies not with the cable system but with the sensors.[10] If they are not sensitive enough, no amount of cable connections will make the system effective; if they are too sensitive, false alarms will be a constant problem. Furthermore, it may take many sensors to protect a single home or building. Even if they were reasonably priced, their installation would likely cost enough to make it worth installing them only in new structures as they are built. In sum, before any fire or burglar alarm system is installed, the adequacy and cost of the sensors and their installation should be tested thoroughly.[11]

Agency Communications

A third application of cable is in the inter- and intra-agency communications of public safety agencies such as the police and fire departments. This application would not be a direct service to the public, and might call for special cable channels unavailable to the public. To begin with perhaps the most striking idea, many cities have considered establishing a *municipal command and control center*. The center would bring together all information on emergency and security services and coordinate service activities, especially during emergencies such as riots. There has been little experience with such centers under any circumstances, much less with cable TV. The centers are likely to run into difficulties unrelated to cable: they are costly to establish and may be inflexible in their use, and it may be managerially impossible to coordinate several emergency services. For instance, it has usually been true in the past that forces from one department resent being commanded by forces from another.[12]

A second internal use of cable TV is for *inservice training* of security forces. Firemen and policemen, for example, could view special and routine inservice programs while on duty in the field, and not have to waste travel time assembling in a single location. Training programs on cable have the added advantage of being able

[9] Ibid., pp. 29-30.

[10] For a review of the various home security devices urban dwellers have used, see George Alexander, "A Nervous New Yorker's Guide to Safety Devices," *New York Magazine*, October 5, 1970.

[11] For a recent study of fire detecting devices, see R. D. Doctor et al., *An Early Detection and Warning System for Fires in Buildings*, The Rand Corporation, R-880-NYC, December 1971.

[12] The issue of agency coordination has already been raised in relation to cable television. See Nicholas Johnson, *How to Talk Back to Your Television Set*, Little, Brown, Boston, 1967, p. 141.

to present action-oriented demonstrations that could not be shown through other media. Fire and police department training has been one of the most consistent municipal service applications on WNYC-TV.

A third internal function of cable is *daily communication* with the agency. For instance, cable TV could be used to call men onto duty, locate cars in the field, or transmit orders of the day. Of these specific functions, the first two probably require only radio, not cable TV, although the radio frequencies could be provided by the FM bandwidth that can be part of every cable system. The transmission of orders of the day can make good use of cable TV, however, since they can include the transmission of photographs of fugitives, missing persons, and other visual records.[13]

Public Safety Education

The fourth application of cable TV in public safety is the broadcasting of public safety information to residents. The information can be linked with specific emergencies, such as announcements of hurricane warnings, tornadoes, or even civil disorders, or they can be linked with the routine showing of programs to improve public safety habits. For instance, police and fire departments might use cable to demonstrate how a homeowner can reduce fire hazards or vulnerability to burglaries, and to convey other helpful hints regarding safety in the home. Community relations and other programs aimed at informing the public about the activities of the police and fire departments might also be developed.

Public safety education programs and internal training programs appear to be the most relevant public safety applications for immediate use on the single municipal channel. For these programs, most municipalities will probably want to use a combination of locally originated programs and prepackaged material (e.g., rented films available through sources well known to most police and fire departments, such as the International Chiefs of Police, Washington, D.C., and the National Fire Protection Association, Boston, Massachusetts).

CRIMINAL JUSTICE ADMINISTRATION

Five applications of cable TV to criminal justice are discussed below: courts processing, security in correctional facilities, training and rehabilitation programs for prisoners, legal services training for lawyers, and public legal information. Courts processing and security in correctional facilities would require several channels with two-way audiovisual capacity, and would not be accessible to the public. Training programs for prisoners and lawyers would not be of much interest to the public. Public legal information would interest a broader audience and could be used on a single municipal channel.

[13] This third use of cable TV is discussed more fully in Mason et al., op. cit.

Courts Processing

In the administration of criminal justice, one proposal has been to use cable TV in connection with prearraignment and arraignment processing in the criminal courts. Dedicated two-way channels could be used, for example, for communication among the judge, the arresting officer, and the arrestee when they are separately located. No such system has been designed or implemented yet, however. Such a use of cable could save a great deal of travel time in cities like New York, where traffic is slow and a large geographic area must be covered.[14] Cable could also be used to transmit fingerprints and other records in conjunction with the arraignment procedure (with the reservation that the fine picture resolution necessary with fingerprints may require special television equipment). Even with better resolution, television reproduction may not be as good as photographs transmitted through the mail, which would be the preferable medium as long as the one- or two-day delay could be tolerated.[15]

The major theoretical advantage of cable TV for courts processing would be the travel time saved. Recent estimates suggest that many police man-hours **are** wasted in travelling to and awaiting courts processing; cable TV could avoid these delays, improve scheduling patterns, and possibly improve the quality of justice by reducing prearraignment and arraignment delays. At the same time, the costs of installing a cable TV system for this purpose have to be weighed against the actual magnitude of the processing problem. Cable may not be justifiable in cities where fewer people are arraigned, delays are not excessive, and lost police man-hours cannot be put to better use. One final matter to be resolved is a legal issue: lawyers and legislators would have to establish whether electronic images on a television set can substitute for personal court appearances in the arraignment procedure.

Security in Correctional Facilities

A small, closed-circuit TV system could be installed in prisons, to enhance security and the safety of prisoners. The claim is that television monitoring could improve the effectiveness of security operations, and perhaps even result in a reduction of guard-hours required.[16] TV surveillance might even reduce suicide attempts and deter guards from abusing prisoners and prisoners each other.

This system resembles any other 24-hour surveillance system, and raises the same issues of privacy, development of effective cameras, excessive waste of channel space, and the human difficulties of constantly monitoring many television screens. One difference is that it would operate only within the prison, probably on a closed-circuit system, and therefore would not demand the channel space street surveillance would require.

[14] Paul Dickstein, "Decentralized Court Processing Over Closed Circuit Television," New York City Bureau of the Budget, January 20, 1971.

[15] J. J. O'Neill, The MITRE Corporation, personal communication, June 1972.

[16] Michael Liechenstein, "Potential Applications of Technology in Correctional Agency Operations and Programs," The New York City–Rand Institute, October 1969, unpublished paper.

Training and Rehabilitation Programs for Prisoners

It has also been suggested that cable TV could provide training and other rehabilitation programs to people still in prison.[17] Prisons offer few rehabilitation programs at present, but the imprisonment period might be an opportune time to provide instruction through specially designed television programs. Prisoners could acquire knowledge about their societal environment and about training opportunities, and also enjoy good entertainment. Although the state of knowledge about effective rehabilitation programs is not well developed, some films on the subject can be purchased or rented (e.g., from the National Film Board of Canada, and New York City).

Legal Services Training for Lawyers

Lawyers share a common body of knowledge and could benefit from special programs televised into their offices or homes. They could acquire information about important cases, with interpretations of new laws, and could exchange information and experiences with other lawyers. The programs could even be made accessible to ordinary viewers, if they are interested in learning more about the legal scene.[18]

Such programs could be a boon to the legal profession. First, lawyers would save travel time by viewing the programs in their homes or offices. Second, if there is enough time in the television schedule, the programs could be shown repeatedly so that a lawyer could view them at his own convenience. Third, if a more sophisticated cable system were installed, each lawyer might even be able to request specific programs to be shown at specific times, in effect using the programs as a video tape library. It is again necessary, of course, to determine whether this cable service will be fully used. If telephone or radio were sufficient for such programs, cable TV would not be a necessity.

Public Legal Information

The final criminal justice application to be discussed is informing the public about the availability of legal services, dispensing legal advice, and publicizing the operations of the criminal justice system. This application seems to be the most likely candidate for immediate use on the single municipal channel. Information on legal services and legal advice might be conveyed through a combination of prepackaged material and locally originated programming. Legal experts in the municipal agencies are likely to know of the most up-to-date sources for programming material. Programs on the criminal justice system could include question and answer dialogs and coverage of court sessions. The latter would make especially good use of the video medium, and would help reveal a portion of public life with which most people are unfamiliar. To date, however, no such systematic coverage of court activities has been carried out. Each court must approve such a procedure, and it is not clear what sessions would be the most informative (e.g., small claims court or divorce

---

[17] Cranberg, op. cit., pp. 35-36.

[18] Monroe Price and John Wicklein, *Cable Television: A Guide for Citizen Action*, Pilgrim Press, Philadelphia, 1972, pp. 22-23.

courts, etc.), or how frequently they should be covered. Without prior experience to rely on, cities may have to adopt a test and demonstration approach at the outset.

ENVIRONMENTAL MONITORING AND CONTROL

Four general applications of environmental monitoring and control are discussed here: monitoring and control of utilities, transportation, pollution, and sanitation. All tend to share similar characteristics with regard to the potential use of cable TV. First, the major limitations of the systems are likely to be a function of the type and number of sensing devices available. Good sensors may be expensive and difficult to install. Second, each system may not really require the use of the video medium, with telephone and radio again serving as distinct alternatives to cable TV. Third, the systems will generally be part of a municipal agency's internal operations, and will not directly involve the public. Fourth, the systems do not make large demands on the information capacity of the cable, but nevertheless do involve a two-way capability. For all these reasons, none of the four applications are likely candidates for the single municipal channel. Moreover, their feasibility on a multi-channel system will still depend on further demonstration, primarily to test the adequacy of the sensors.

Utilities Monitoring and Control

Fire and burglar alarm systems, while discussed above under public safety, also exemplify a whole class of environmental monitoring systems that could be created with cable TV. Gas, electric, and water meters, for instance, could be read by cable TV instead of meter-readers going from house to house. This may encourage utilities companies to cooperate with the municipal government in developing the most desirable system. The meter-reading system could use the same connections and operate on the same channels as the fire and burglar alarm systems. Cable could also monitor water levels and the capacities of other central municipal facilities, with response mechanisms built in that would initiate corrective actions to prevent floods, blackouts, and other emergencies.

The availability and cost of adequate sensors is again an issue. More important, automated utilities monitoring systems may not save enough money or guarantee penetration into enough homes to warrant being implemented.

Transportation Monitoring and Control

Secondly, cable TV could assist in the development of traffic control and other systems designed to monitor transportation conditions. Street-corner cameras, for instance, could focus on busy intersections, and the timing of signal lights could be adjusted to traffic flow.[19] Computers also can be used to analyze the pattern of traffic at street corners, so that traffic signals can be appropriately timed; the cable can carry the return information that controls the signals. A car counter that does not

[19] Mason et al, op. cit., p. 106.

require a visual image can also serve these functions, however, in which case telephone or radio communication would suffice. Cable TV would be an advantage only if visual information became more critical, or if adequate telephone or radio links were unavailable.

Pollution Monitoring

Cable TV could be used to determine air, noise, and water pollution levels. Sensor devices located at critical points (such as smokestacks) would transmit signals to a central location. The pollution detecting systems could be especially valuable where city governments enjoy good cooperation from their citizens. With cooperation, these cities could encourage people to drive their cars less or burn less trash during excessively polluted periods; industrial plants might also be shut down. Again, the main issue for cable TV would be whether a good sensor system can be developed, and whether the information flow is appreciable enough to demand the use of television, or whether radio or telephone signals would suffice.

Sanitation Monitoring

Any system of street surveillance, whether for crime prevention or for traffic control, could also be used to determine needs for sanitation collection, perhaps on dirty streets that require additional garbage pickups or street sweeping or both.

This system would be free from one of the major difficulties of 24-hour anticrime surveillance: streets would not have to be monitored constantly, since the sanitation department could merely look successively at each street corner, perhaps daily. Little channel space would be used, and only a few persons would be needed to watch the incoming television scenes. Such a system could also improve sanitation collection procedures. At present, most cities do not have daily advance information for determining what streets have how much garbage. Instead, they have had to rely on fixed collection and street sweeping routes. If rapid population turnover at the neighborhood level means constantly changing patterns of the distribution of garbage, then fixed routes might not use sanitation resources as efficiently as would flexible routes based on advance information.

On the other hand, the installation of a cable-TV system for monitoring garbage would have to compared with other ways of gaining the necessary information. For example, citizen volunteers stationed throughout the city could telephone daily reports to the sanitation department. This would be much cheaper than the cable TV system and would not raise the invasion of privacy issue, an issue that any mechanical street surveillance system is bound to encounter.

HEALTH CARE

Four applications of health care are discussed: telemedicine, intra-agency communication (including inservice training), continuing medical education, and public health education. Telemedicine and intra-agency communication would require two-way links and perhaps a large number of television channels. Though the feasibility of telemedicine has already been demonstrated, neither application is likely to be relevant for the single municipal channel now available for public services (with the exception of inservice training). The third application, continuing medical education, resembles legal services training for lawyers. The programs would be aimed at a small audience, but could be carried over the single municipal channel. Unlike legal services training, however, some medical programs might be offensive to a general audience, and somehow would have to be shown to doctors only. The fourth application, public health education, would be well suited for immediate use on the single municipal channel.

Telemedicine

The concept of telemedicine is simple: the patient and medical assistance are at different locations, and communicate by telephone, two-way radio, or television. In recent years, telemedicine has become increasingly associated with a television link. This is a promising use of cable TV, for two reasons. First, the current geographic distribution of medical personnel and specialized facilities in the United States does not match well with the areas of health care need. For instance, few doctors work in inner city ghettos.[20] Telemedicine could help to compensate for the geographic mismatch, bringing specialized medical care into ghettos, nursing homes, and other isolated communities.[21] Second, television can transmit an abundance of information that other media cannot provide. For instance, patient-doctor interactions (teleconsultation) are richer on two-way television than on telephone, and rapid communication of X-rays, electrocardiograms, and other records (telediagnosis) flatly require the wideband communication capability of cable TV.

Several demonstrations of telemedicine have already been carried out. Perhaps the most publicized is the television link between Massachusetts General Hospital and two remote medical facilities, one at Logan Airport in Boston and the other at a Veterans Hospital in Bedford, Massachusetts. By 1971, doctors at Massachusetts General had already aided in the counseling and treatment of over two thousand patients, for both medical and psychiatric services.[22] (An interesting phenomenon that emerged was that patients reported the psychiatric interaction over two-way television to be more comfortable and relaxing than face-to-face interaction.[23]

Municipalities considering new cable TV systems would have to plan much beyond the current single municipal channel to create similar telemedicine links. The links require two-way capabilities, several pieces of medical equipment at the various service locations, and several television channels. For instance, depending on the number and scheduling of outlying service stations, two or three channels might have to be set aside in case of emergencies or heavy demand. Ultimately, the links might provide improved health care for ghetto residents and for other special

[20] James W. David, "Decentralization, Citizen Participation, and Ghetto Health Care," *American Behavioral Scientist,* Vol. 15, September-October 1971, pp. 94-107.

[21] *Committee on Telecommunications,* op. cit., pp. 63-64.

[22] Konrad K. Kalba, "Communicable Medicine: Cable Television and Health Services," report prepared for the Sloan Commission on Cable Communications, September 1971, pp. 40-42.

[23] Kenneth T. Bird and M. E. Kerrigan, "Telemedicine: A New Health Information Exchange System," paper presented at the Medical Services Conference of the American Medical Association, May 1970.

communities such as nursing homes, juvenile homes, prisons, and mental hospitals. Special communities tend to occur in isolated parts of a metropolitan area far from comprehensive medical facilities, and they all serve relatively immobile populations. Perhaps the most valuable telemedical links, however, will transcend the boundaries of metropolitan areas.

Telemedicine could bring improved health care to rural communities, small towns, and remote communities throughout the country. To serve such communities would require clearly regional, not merely metropolitan, planning and coordination.

Intra-agency Communications

Cable TV seems to have a logical place in intra-agency communications for health care because of the organizational features of most municipal health care systems. Health services generally have the following features: (1) abundant face-to-face contact, both between the patient and medical personnel and among medical personnel; (2) heavy reliance on patient records that must be frequently stored and retrieved; (3) fixed facilities such as hospitals and clinics that require much travel by patients, medical personnel, and visitors; and (4) a great deal of functional specialization, requiring, for instance, patient travel within a hospital for different tests, and communications between laboratory and clinic, between medical services, and among administrative services. These organizational features suggest that mechanical communication, such as cable TV or telephone, could yield significant savings in time and money.

The most likely immediate application for cable TV, however, will be inservice training programs on the single municipal channel for public health officials.

Continuing Medical Education

As a special type of inservice training, and as with lawyers, cable TV can provide continuing medical education for doctors. Both scientific breakthroughs and medical advances continually render portions of medical knowledge and practice obsolete, and doctors, especially general practitioners, are hard pressed to keep abreast of their fields.

Continuing medical education might use the visual medium to better advantage than would legal services training, because it relies so heavily on visual demonstrations, if not actual clinical practice. Existing prototypic medical programs support this proposition. In one experiment, Community Medical Cablecasting (New York City) telecasts programs exclusively to doctors in twelve CATV systems. Local medical officials will know of other program sources (e.g., the Network for Continuing Medical Education, New York City).[24] Programs on many specialized subjects could be developed and continually revised if necessary. They could be shown at specific hours or, preferably, at each doctor's convenience. Though this latter arrangement

---

[24] Rudy Bretz, "The Application of Cable to Continuing Medical Education," Paper 8 in L. L. Johnson et al., *Cable Communications in the Dayton Miami Valley: Basic Report*, The Rand Corporation, R-943-KF/FF, January 1972; and Kalba, op. cit., pp. 11-20.

would be expensive to install, the doctor could draw on the programs as part of a video tape library, requesting that they be shown in his office or home at the hour he wants to view them; he would not have to travel to see them. Similar arrangements could be made for nurses, pathologists, psychiatrists, and other medical specialists.

Unlike legal services programs, medical programs would probably not be made available to the ordinary home viewer. Being highly technical, a program might not even interest medical people in another specialty, much less the lay public. The programs might also offend the taste of the ordinary viewer. For these reasons, it is likely that continuing medical education will require television channels strictly dedicated for such use, or some other arrangement for limiting the programs to the target audience.

Public Health Education

Unlike telemedicine, educational programs in public health would require only one-way television transmission, much like today's conventional television. They could range from actual demonstrations of home medical care to information about the local health care system and how to use its facilities.[25] One research report has pointed out a case in which a dental health program was broadcast over local television, successfully conveying information about dental health where pamphlets had failed in the past.[26] It made good use of the television medium, since it demonstrated important procedures that called for moving visual images. On information about local health care systems, one writer has poignantly described the need for such information by suggesting that "a person visiting Europe can get better advance information than a person entering the 'foreign country' of the hospital."[27]

Public health education on cable TV could improve many aspects of health care. First, improved information might enable people to reduce the time they spend traveling to health facilities and being referred from one office to another. Good information could direct them to the correct office in the first place. Second, well-designed demonstrations could lead to better health care practices by the ordinary citizen. Third, the programs could inform the public of the latest medical advances and treatments available, and of public health conditions peculiar to a given city or even a given neighborhood.

As with other areas of public education, success here will hinge on program style and content. Poor production can make them uninteresting and they will not attract audiences. Currently, municipal officials may contact any number of distributors of public health films and video tapes (e.g., National Education Television, New York City). Local public health officials are likely to know these sources and can also determine how appropriate such prepackaged materials will be for their own communities.

[25] Kalba, op. cit., pp. 25-31.
[26] Price and Wicklein, op. cit., p. 22.
[27] From Michael Crichton, *Five Patients: The Hospital Explained,* as quoted by Kalba, op. cit., p. 27.

SOCIAL SERVICES

The scope of social services is difficult to define, as it varies from city to city.[28] The services can be defined, in a negative sense, as those services not normally considered part of health, education, sanitation, and other municipal functions. Or they can be defined in a positive sense, as those services that provide help to people who are disadvantaged, with the definition of "disadvantaged" being construed broadly. It should include the blind, the deaf, the handicapped, the aged, and the poor; but it should also include people who may have only a short-term need for help—a family needing temporary shelter because of a fire in its home, an office worker temporarily out of a job and needing information about new opportunities, consumers victimized by unfair price schemes. Thus defined, social services encompass many functions. Some are not even administered directly by municipal governments, but by private agencies, but they constitute a major and increasing portion of municipal government activity.

Four general applications of cable to social services are discussed here: extending person-to-person services, intra-agency communications (including inservice training), community centers, and social service information. The first three applications will usually require a two-way capability; the only applications likely to be of immediate relevance for the single municipal channel are inservice training for municipal employees and the dissemination of social service information to the public.

Intra-Agency Communications

In some cities, the administrative apparatus for social services is already so large and complex that cable TV potentially could bring the same types of improvement in communication as might be possible in health care administration. To this extent, the organizational features previously mentioned in relation to health care administration would probably have similar importance in social service administration. One can well imagine saving clients' time in travelling to and waiting at district offices, improving records retrieval, and improving intra-agency communications.

All these applications would require a multichannel capacity as well as two-way capabilities. They must also be field-tested for feasibility even if they become technically possible.

Extending Person-to-Person Services

With certain social services, cable TV might actually be used to improve the services themselves, not merely provide better information about them. Most of these uses would require two-way links between individual homes and social service or other municipal offices.

---

[28] For a broad review of social services and cable TV, see Robert O'Brien, "Cable Communications and Social Services," report prepared for the Sloan Commission on Cable Communications, June 1971; and idem, "1971-1973 HRA Telecommunications," New York City Human Resources Administration, September 1971.

Many small groups of people in every city would be well served by two-way televised social services. Most of these people are confined to their homes, some because of physical handicaps, others because of age (extremely old or young), and yet others for reasons of safety, penury, or sickness. Televised personal services could improve their well-being. For instance, one experiment has already been conducted to see whether handicapped children confined to the home can carry out some portion of their studies by two-way television.[29] More speculatively, public health nurses, social workers, and other aides could "visit" the home by television and consult the home-bound person. Consultation could be made available at hours or to neighborhoods that are not now served by personal visits. Many routine nurse's visits for pre- and post-natal care, for instance, have been eliminated because of the potential danger to the nurses in visiting a particular neighborhood, especially at night.[30]

At least two problems are associated with this use of cable TV. First, the people who need help the most may not subscribe to cable TV at all without subsidy or preferential rates, and they may live in those very neighborhoods that will be integrated last into a municipal cable TV system.[31] In other words, they may be as difficult to reach as they ever were. Second, there may be an undesired second-order effect if a successful program of this nature is established. This is that caseworkers (whether social workers, nurses, or others) will gradually use television as a substitute for actual home visits for all their clients. Such a situation may be undesirable because it will steadily diminish the caseworkers' knowledge of their clients' unique physical and social environment, a knowledge that can be gained only through personal visits.

Neighborhood Centers

One service of special importance in the future is the development of local neighborhood centers for all phases of social services. The centers might be a more feasible immediate application for cable TV than extensive connections into individual homes, since they would cost less and would lend themselves to experiment. They could supply a great deal of social service information and, with appropriate terminal equipment, could distribute municipal and federal forms as well, saving people from traveling to City Hall. Computer programs could even help people fill out their tax returns.

The centers could be placed in every small neighborhood, and become the focal point for social activities. The centers could be especially effective in the suburbs, where municipal offices may not even exist and travel into the central city is not

[29] Boyce Rensberger, "Cable TV: 2-Way Teaching Aid," *The New York Times,* July 2, 1971, p. 16.

[30] Some neighborhoods are deemed so unsafe that mailmen even refuse to enter specific houses, whose residents must go to the post office for their mail. See Rosa Guy, "Black Perspective: On Harlem's State of Mind," *New York Times Magazine,* April 16, 1972.

[31] Preferential rates for low-income groups and other disadvantaged people can be specified in the cable franchise. See Baer, op. cit., pp. 62, 107. Such rates can also be justified on economic grounds, since it costs less to wire inner-city areas of high residential density than to wire suburbs; see Robert K. Yin, in Baer et al., op. cit. "Cable Subscription Fees" in chapter 4, Sec. III. The FCC has also warned that "a plan that would bring cable only to the more affluent parts of a city, ignoring the poorer areas, simply could not stand." See Rivkin, op. cit., p. 68.

convenient. It is possible that the federal government will sponsor demonstration tests of neighborhood centers in the near future, in which case municipal governments can hope to gain early feedback on their feasibility and cost.

Social Service Information

In the most immediately feasible application for a single municipal channel, cable TV could be used to improve public education about social services. Series of programs might be devoted to drug abuse, employment opportunities, home care and legal advice, information on the use of municipal services, and new regulations. Because cable TV can focus on specific audiences, these programs could eventually cater to the needs and concerns of particular communities, whether they are inner-city ghettos, ethnic villages, or suburbs.[32]

Dayton, Ohio, has already established an attractive precedent. There, the municipal ombudsman appears at frequent intervals on over-the-air television, covering a few of his more interesting cases with site visits and other audiovisual materials.[33] Other popular programs have dealt with the problems of drug and alcohol abuse.

As with public health education, the key to successful social service information programs will probably be program style and content. The telecasting of job opportunities, for instance, will be very dull if it does no more than recite lists of job openings and their related details. It will be more interesting and informative if it also includes short films and demonstrations of the types of jobs available. (Once again, local social service professionals will know about sources of prepackaged program materials, but can obtain additional information from the Cable Television Information Center, Washington, D.C.) Other than program content and style, there appear to be few barriers to the use of cable TV for disseminating social service information. Since social services are so dependent on communications—where to go, whom to see, what to ask for—cable TV is potentially valuable, as long as programs are attractive and are oriented toward clients' perceptions and needs.

REGIONAL COMMUNICATION

Although many of the new urban cable systems may not incorporate an entire metropolitan region, two general applications of regional communication will be discussed for illustrative purposes: inter-agency communication, and fascimile mail. Both require a regional cable TV network and two-way multichannel capabilities. For these reasons, neither these nor other regional applications are likely to be amenable to a single municipal channel.

[32] For a discussion of the types of information that might be useful to a ghetto community, see Herbert S. Dordick et al., *Telecommunications in Urban Development,* The Rand Corporation, RM-6069-RC, July 1969. The authors describe a survey made in Los Angeles and New Orleans indicating that the failure to circulate community information within the ghetto and between the ghetto and neighboring communities is largely responsible for the isolation of ghetto residents and for their inability to enter into the economic mainstream.

[33] Robert K. Yin, "Cable Television and Public Interest Programs," Paper 5 in L. L. Johnson et al., op. cit.

Interagency Communication

On a metropolitan area basis, considerable communication among officials of different agencies and different levels of government may be desirable. For instance, police departments often express the desire to coordinate their activities, especially when pursuing fugitives who cross agency boundaries.

Cable TV would facilitate such communications by (a) broadcasting area-wide information quickly, (b) serving as a communication link during area-wide emergencies, (c) substituting for face-to-face communication on many routine matters, thereby reducing or eliminating travel time, and (d) providing in-service training to officials.

At present, evidence is scanty on these uses of cable TV, but the New York Metropolitan Regional Council has begun a feasibility study for agencies in the New York area.[34] Cable TV seems to be an excellent way to overcome New York's excessive travel distances and times and its frequent difficulties with telephone communication. The feasibility study will identify the communication patterns among the agencies in the New York area, and specify where cable TV could be most effective.

Britain has been exploring the use of television for intercity conferences. Any group of people can schedule an appearance in a studio conference room in one city and communicate by television with a similar studio in another city.[35] How popular these studios will be, or how such facilities compare with alternatives such as conference telephone calls, remains to be seen.

In summary, cable TV presents many possibilities for improving both vertical and horizontal communications among offices. Vertical communications would link city, state, and federal agencies concerned with the administration of similar programs, or that already administer joint programs. Horizontal communications would link agencies dealing with different programs (e.g., housing, sanitation, and health) or dealing with similar progams in different parts of a region (e.g., the police departments in a metropolitan area). In either vertical or horizontal communication, cable TV could allow officials of one agency to learn more about and work more closely with officials in another agency.

Facsimile Mail

On a regional basis, cable TV has also been promoted very speculatively as a substitute for communications presently conducted by mail.[36] One possibility would be to transmit documents on a television channel and reproduce them at the receiving end with a Polaroid camera. This very rapid system would avoid the one- or two-day delay of the mails. It would also transmit documents faster and make more accurate reproductions than do current teletype links that use telephone lines. A second possibility would be to transmit documents on a television channel and have

[34] Daniel J. Alesch, *Intergovernmental Communication in the New York-New Jersey-Connecticut Metropolitan Region*, The Rand Corporation, R-977-MRC, May 1972.

[35] Richard C. Harkness, "Communications Innovations, Urban Form, and Travel Demand," Urban Transportation Program, University of Washington, Research Report 71-2, January 1972. This report also has a comprehensive bibliography on the communications aspects of cable TV.

[36] This and other interactive uses of cable television are discussed in Walter S. Baer, *Interactive Television: Prospects for Two-Way Services on Cable*, The Rand Corporation, R-888-MF, November 1971, pp. 21ff.

them read but not reproduced. This might cut down on paper consumption and even reduce the amount of trash to be disposed of.[37]

The use of cable TV as a substitute for mailed documents has some attraction if projected mail levels are anywhere near accurate. One source has predicted that by 1980 there will be 108 billion pieces of mail annually and the volume of paperwork will have doubled.[38] Much of this flow consists of bills, payments, and other financial statements that involve only a transaction and do not necessarily require the transmission of a piece of paper. On the other hand, the use of cable TV will be more expensive than the mail system; to be effective, cable TV terminals would have to saturate a large number of offices or homes. The key question is likely to be whether the one- or two-day savings in time compensates for cable's higher cost.

[37] Edward H. Blum, *The Community Information Utility and Municipal Services,* The Rand Corporation, P-4781, February 1972.

[38] "The Future of Broadband Communications," October 1969, as cited by Ralph Lee Smith, *The Wired Nation,* Harper and Row, New York, 1972. Originally published in shorter form in *The Nation,* May 18, 1970.

V. LESSONS FOR MUNICIPAL DECISIONMAKING

Cable TV offers a wide variety of uses for municipal services. This review has described many of the applications conceived to date, covering public safety, criminal justice administration, environmental monitoring, health care, social services, and regional communication. For municipal officials engaged in franchise negotiations, however, what are the relevant issues for decisionmaking?

First, *for using the single municipal channel presently allocated by the FCC, municipal officials should consider the several variations possible in site and program selection.* Some programs may be better administered through neighborhood centers, for instance, than through home delivery if the cable system fails to penetrate homes of low-income families. In addition, programs should be introduced incrementally, with the programming day only slowly expanded.

Second, *no matter how the single channel is administered, its operation will require an additional outlay of funds.* The costs will vary from city to city, depending on the degree of coverage and type of equipment used; but whatever the city, cable TV will require an outlay of funds above current operating levels. Only a few cable TV applications, such as prison surveillance systems, have any hope of realizing long-term monetary savings. Given the general financial plight of the central cities, and especially given the fact that some cities have recently had to reduce municipal services, not expand them, support for cable TV applications must be weighed seriously against alternative expenditures.

Third, *the single channel will accommodate municipal applications only of the traditional broadcasting variety,* and hence will not constitute a good test of the usefulness of full, two-way systems. The single channel will be operated as a broadcast facility, while the two-way system will be operated as a communication facility, with many different applications. Two-way systems are technically feasible, but the current cost of cameras and other terminal equipment is still too prohibitive to be installed in large numbers in the near future.[1] To this extent, the initial use of cable TV for municipal services will be limited to one-way applications. Other applications, such as telemedicine and televised visits into the home by caseworkers, will not be possible until cable TV systems have two-way video and audio communications. Thus, the time horizons for adopting the various municipal applications of cable TV are different.

[1] Baer, *Interactive Television,* p. 36.

Fourth, *only a few of the applications have been field tested.* Many basic cable design features are also untested. For instance, any system with a large number of cameras scattered about a city for two-way audio-video feedback will also pick up considerable amounts of noise. Every additional camera, even if not in use, can potentially add extraneous signals, such as might as might come from a ham radio operator and the two-way radio connections of police departments, taxi companies, and other mobile units. Field demonstrations are needed to determine noise levels and how much noise is tolerable. Field demonstrations are also needed to show the service gains and social usefulness related to each of the applications. Thus far, there is very little evidence along these lines, either.

Fifth, *many of the service changes resulting from cable TV may not lead to dramatic improvements in service.* In most applications, it is true that cable TV can improve service communications, reduce travel, and make services more accessible. But improved communication may improve services only marginally or not at all.

An example is a recent experience with neighborhood storefronts in New York City.[2] The storefronts were established to "bring government closer to the people" by having city representatives in the neighborhood. Residents were encouraged to bring service complaints to the storefront, with the idea that the storefront could effectively dispose of the complaints, since they did not typically deal with problems falling within the purview of a single agency. Often, as in the case of exterminating rats, the work of several agencies had to be properly coordinated, and no municipal coordinating agency existed—nor did the storefront prove to be a successful substitute. Even when complaints were the responsibility of a single agency, the storefront was often not much more effective in obtaining agency action than any ordinary citizen would have been. Since success was rare, the storefronts tended to make themselves less available to residents, e.g., by moving to noncentral locations, operating at odd hours, or appearing to be very busy during regular hours.

The behavior of the storefronts seems entirely adaptive and normal. There was no reason to raise people's expectations by accepting many complaints that could not be dealt with. The behavior illustrates how communications alone cannot overcome basic service shortcomings, and how in fact the reverse can occur: where municipal programs are unresponsive, there is a natural tendency to *reduce* communications about them.

In the same vein, municipal officials should not assume that because cable TV is a new technology it can magically solve a city's most pressing problems. Such thinking would be an unfortunate extension of "tech-fix," the supposition that science and engineering alone, if given the funding, have the conceptual wherewithal to remedy social ills.[3] On the contrary, the most effective cable TV system is likely to require heavy participation by municipal officials and community leaders, and to call on their knowledge in designing and operating the cable TV system. For instance, cable TV can be used to create either a highly centralized or highly decentralized administration of services.[4] Only the officials and residents of a city can decide whether a major reorganization of services is desirable (decentralization is

---

[2] Douglas Yates, The New York City–Rand Institute, personal communication, September 1970.

[3] H. L. Nieburg, "The Tech-Fix and the City," in Henry J. Schmandt and Warner Bloomberg, Jr. (eds.), *The Quality of Urban Life,* Sage, Beverly Hills, California, 1969, pp. 211-243.

[4] Kalba, op. cit., p. 35.

the current vogue), and how the cable TV system is to be used. Cable TV by itself, in other words, will play only a passive and secondary role in any such major reorganization.

Sixth, *to be used effectively for several municipal services, cable TV will have to reach those segments of the population that have traditionally been the most difficult to reach.* Some of the most important service applications of cable TV depend on reaching people and families who are not now reached by municipal services. As previously noted in the discussion on social services, these groups may include the aged, the poor, and racial and ethnic communities. Existing experience with cable TV suggests that it takes special efforts to enroll these groups as cable subscribers. Otherwise, subscribers will tend to be middle- and upper-middle-income residents, perhaps following a pattern similar to color television ownership. According to a survey of one city, about one-third of all families with incomes of $5000 or less owned color television sets, whereas three-fourths of the families with incomes of over $15,000 owned such sets.[5] Furthermore, because incomes are usually higher in the suburbs than in the central city, color television ownership is also much more a suburban phenomenon. If the analogy between color television and cable TV is appropriate (and it may be, since color sets are usually purchased on the installment plan, and the payments can be considered similar to the monthly subscription fees for cable TV), cable TV penetration will also be dominated by higher-income suburban residents.[6] Obviously, if a municipal cable TV system fails to penetrate low-income groups and neighborhoods whose populations have the highest rates of disease, illness, fires, or crimes, it will not improve municipal services geared to these problems.

Seventh, *many cable TV applications can also be performed through other communications media, including face-to-face contact, the telephone, two-way radio, and the mails.* This is particularly true of simple signal systems like burglar or fire alarms, where the full bandwidth afforded by television is not needed. Similarly, many applications require only voice communications, and telephone or two-way mobile radio may be sufficient. Finally, some applications, such as sanitation monitoring, can even be accomplished through a network of resident observers or sanitation inspectors, and do not require an electronic surveillance system. Any consideration of the potential uses of cable television must therefore include an analysis not only of cable's feasibility, but of its advantages over other media. The failure to make such an analysis has been an important shortcoming of almost all previous discussions of cable television.

For instance, most cable communications require that the transmitting and receiving terminals be geographically fixed. This is not true of radio, telephone, and the mails, all highly flexible as to geographic location. In face-to-face communication, the two people of course must be in the same place at the same time. Geography therefore affects the desirability of all these media for various functions.

Similarly, these media can be compared by their temporal characteristics. Cable TV, radio, and telephone communications are the most flexible; all can be used for both simultaneous and delayed communication. For delayed communication, messages can be recorded and stored on tape. Mail, on the other hand, involves a

[5] Robert K. Yin, "Cable on the Public Mind," *Yale Review of Law and Social Action,* Vol. 2, Spring 1972, pp. 289-297.

[6] But see footnote 31 above.

built-in time delay, while face-to-face contact has a somewhat opposite limitation: it cannot provide delayed communication.

A third important characteristic is the amount of information the various media can communicate. Face-to-face interaction is the richest from this standpoint. Cable TV also permits two-way audiovisual communication, though the camera is not as perceptive as the eye, and other attributes of face-to-face interaction (mood, gesture) may be lost when televised. Cable TV is not nearly as restrictive as radio, telephone, and mail, however. The radio and telephone, while interactive, are limited to audio messages. Mail is limited to written communication, and is not immediately interactive.

These are but three of the more important ways in which the five media are to be compared. Many other attributes could be investigated, such as the necessity for capital investment, costs of operation, susceptibility to distortion or interference, and reliability of performance.

Eighth, *the use of cable TV in municipal services will raise controversial social issues*. Invasion of privacy is perhaps the most important, as discussed earlier. In addition, the control of cable TV may even be a factor in the balance of community power.[7] For instance, *street* surveillance systems will mean a large increase of information flowing to the municipal agencies. Agencies may thus have more control over street activities, potentially approaching Orwell's nightmare vision of Big Brother monitoring people's behavior by television. On the other hand, cable TV could be used for *agency* surveillance systems, allowing residents to view the activities, say, inside a police precinct house, a schoolroom, a courtroom, or the emergency room of a hospital. In this way, more information would flow in the opposite direction—to the residents, who might then be able to make agencies more responsive to public needs, and have a new source of power over their municipal government.

Municipal officials should be prepared to deal openly with all social issues raised. Open discussions may help to allay some undue fears, but officials should expect that some compromises may have to be made between agency-oriented uses of cable TV and consumer or citizen desires.

[7] Nicholas Johnson, op. cit., pp. 140-141.

Appendix

WNYC-TV AND ITS IMPLICATIONS FOR MUNICIPAL
CABLE TELEVISION STATIONS

New York City has the only municipally owned and operated noncommercial public television station, WNYC-TV, Channel 31. It began in 1961 as an FCC experimental UHF television station to provide training for civil service employees. Courses for firemen, policemen, nurses, and clerks constituted the basic program format. In addition, since most television sets available at that time were not equipped to received UHF signals, the early programming content did not reflect the interest of the public viewer.

The station's history and experience should interest cable operators and municipal managers in cities where the FCC requires new cable systems to furnish a municipal channel for use by local government. By linking civic agencies to a central television studio used for municipal cablecasts, the cable operator could provide services well beyond those envisioned by New York's municipal channel. The challenge lies in maximizing the talents of available creative personnel, and in encouraging agencies to pool their equipment, talent, and budgets, to use one comprehensive television center. This brief appendix describes WNYC-TV's recent experiences and the relevant lessons for urban cable systems in the immediate future.

HISTORY

The Municipal Broadcasting System, which includes WNYC-TV and an AM and FM radio station, was conceived as a public service activity of the Mayor's Office. The System has operated for many years (in fact, the radio station went on the air on July 7, 1924), but two recent events have had a great influence on its operations, particularly on the television component. First, from 1961 (the first year of WNYC-TV's operation) to 1969, WNYC-TV received sufficient public attention, equipment, and personnel, partly as a result of its home in City Hall. In 1969, however, the Mayor reorganized several major agencies, and the Municipal Broadcasting System was moved out of the Mayor's Office and into another agency, the Municipal Services Administration. This change in organizational location dramatically altered the administration of the System, as the new agency environment has been oriented

113

toward formalized municipal procedures poorly suited for operating a seven-day-a-week, year-round television studio. To take just one example, the organizational relocation also involved a physical relocation, and the new physical environment does not include proper air conditioning for the studio or studio equipment.

The second change came with the city-wide budget cutbacks in June 1971. WNYC-TV's staff was cut by 25 percent, and the station's broadcast day had to be shortened. Although a job freeze subsequently went into effect, and the station had to look for volunteers and trainees for staffing, the station continues to offer its viewers 11 hours a day of programming during the week, and 5-1/2 hours on weekends.

These two changes have made it all the more difficult for the station to encourage the use of television by the municipal agencies. The station does alert them to the availability of air time, and the program director has especially suggested WNYC-TV as a means of servicing the needs of the Spanish-speaking population. The two agencies that have most consistently used the station are the Fire Department and Police Department. Interest within these agencies is high, especially because inservice training by television has become a standard practice for both. The Police Academy has its own television facility for in-house training, and the Police Department has used its air time to inform officers of new regulatory changes, such as in the penal law.

Other agency users have included the Department of Social Services and the Department of Hospitals, under the aegis of the Commissioner of Health. But the overall agency demands for air time have never created a scheduling problem.

RESOURCES

WNYC-TV presently has an operating budget of about $700,000, with a full-time staff of about 40 persons. The station has no funds to purchase programming, and all programs except the news are broadcast on videotape.

The studio equipment, listed in Table 2, combines the modern and the old. It is 80 percent functional, and there is no preventive maintenance or backup equipment. Like many other stations, WNYC-TV relies on its technicians to repair faulty parts, or tries to obtain replacements. There are about 18 full-time technicians for operating the studio; many are retired from the military, having worked for the Army Pictorial Service or Army Communications. Others are young people who began as volunteers and were able to secure job appointments. Because the station does not have enough full-time technicians, it also relies on volunteers and trainees. The nontechnical staff of the television sector of the Municipal Broadcasting System includes four managers (director of operations, program manager, production manager, engineering chief supervisor), eight clerical staff (stenographer, typist, etc.), three directors, four announcers, two film editors, one publicity assistant, one illustrator, and one production assistant.

These resources being relatively sparse, municipal agencies must develop their own programs for telecasting. WNYC-TV does not have, for instance, a sufficient production staff. An agency wishing to use the facilities of WNYC-TV must call on its own personnel to plan and develop programs and to train its own participants. (Several of the larger agencies have their own television personnel and camera and

Table 2

WNYC-TV'S STUDIO AND EQUIPMENT

<u>Studio:</u>	20 x 40 x 30 ft, L-shaped
<u>Control room:</u>	18 x 25 ft
<u>Equipment:</u>	2 Norelo PC-70 Plumbicon cameras, equipped with 10:1 zoom lenses
	3 RCA TK 60 cameras (black and white cameras of 1966 vintage, used for remote pickups from City Hall)
	3 Ampex VR 1200 tape recorders (2 machines equipped with Editec)
	RCA TS 25 and 11 switchers
	2 RCA TK 27 film and slide chains
	2 RCA TP 66 film and slide chains
	Collins 212 T2 audio panel
	17 microphones (including Sony tietacks, 635, RE 50 and 666 mikes)
	1 Arriflex 16 mm camera
	1 Auricon 16 mm camera
	All incandescent lights, converted to quartz

recording equipment, though in most cases these resouces are considered of secondary priority to the agency's major problems.) All preparation and organization are therefore done before the "cast" arrives in the studio.

PROGRAMS

Figures 3 and 4 are two sample weekly schedules, one from 1970 and the other from 1972, revealing the decrease in programming time. The schedule shows two types of programs: those produced by WNYC-TV exclusively (in italics), and all other programs, including reruns of those produced by the Public Broadcasting Corporation (in roman).

The Police Department, one of the first city agencies to use WNYC-TV, has been a steady "sponsor" since its inception. Its training program, "Around the Clock," is on the air five days a week, and is linked to the training operations in precincts throughout the City. After the program is viewed in the precinct house, there is a follow-up lecture and discussion. The general public can also view the program and gain insight into the operations of the Police Department. Nonetheless, this public exposure does not seem to inhibit the content of the Police Department's training program.

Other agencies, such as the Department of Social Services and the Housing Authority, use WNYC-TV less frequently. Some of these agencies have had to seek

MAY, JUNE : 1970

	Sunday	Monday	Tuesday	Wednesday	Thursday	Friday	Saturday
9:00							
9:30		AROUND THE CLOCK					
10:00		SESAME STREET					
10:30							
11:00		University Broadcast Lab	Consultation	Documentary Hour	All About TV	Staff Meeting on the Air	
11:30		Casper Citron	Community Action				
12:00		NET Journal	NET ½ Hour	It's Fun to Read	NET Hour	NET Festival	
12:30			Compass	Achievement			
1:00		The Book Scene	Staten Island Today	Focus on Books	Your Right to Say It	Manhattan	
1:30		AROUND THE CLOCK					
2:00		Wings to the World	Big Picture	TV Travelogs	Fair Adventure	Documentary Hour	NET Festival
2:30		NASA Presents	Community Report		Continental Comment		
3:00	English for Americans	RETURN TO NURSING			All About TV	All About TV	NET Journal
3:30	Fair Adventure	One to One	Manhattan	Sight and Sound	Lee Graham Interviews		
4:00	Sight and Sound	Forsyte Sage	NET Playhouse	Advocates	The Show	Soul!	NET ½ Hour
4:30	Staten Island Today						Advocates
5:00	Guten Tag	Urban Challenge		Casper Citron	Health Education	Changing Black Community	
5:30		MAKE SURE - MAKE SHORE					NET Hour
6:00	University Broadcast Lab	Community Action	English for Americans	Lee Graham Interviews	It's Fun to Read	Staten Island Today	
6:30		WNYC - TV NEWS					
	British 1/4 Hr	Journey	New Horizon's Dateline		Camera 15	Scope	It. Panorama
7:00	One to One	On the Job	Around the Clock	On the Job	Around the Clock	On the Job	On the Job
7:30	Focus on Books	Community Report	Speaking Freely	The Book Scene	Registered Nurse	Brooklyn College	NET Playhouse
8:00	Brooklyn College	Make Sure - Make Shore		University Broadcast Lab	Rise of the American Nation		
8:30	Your Right to Say It	Consultation	Changing Black Community	All About TV			
9:00	The Book Scene	New York Report			One to One	Sight and Sound	Forsyte Sage
9:30	Documentary Hour	WNYC - TV NEWS					
		British 1/4 Hr	Dateline	New Horizons	It. Panorama	Camera 15	Forsyte Sage
10:00		Brooklyn College	Music Specials	Manhattan	Guten Tag	Urban Challenge	The Show
10:30	With Mayor Lindsay	Eye on the Universe		Eye on the Universe		Eye on the Universe	
11:00							Soul!
11:30							

Fig. 3—Schedule from 1970

JULY, AUGUST : 1972

	Sunday	Monday	Tuesday	Wednesday	Thursday	Friday	Saturday
9:00							
9:30							
10:00							
10:30							
11:00							
11:30							
12:00	THE POLICE COMMISSIONER						
12:30	AROUND THE CLOCK						
1:00	SESAME STREET						
1:30							
2:00	MR. ROGERS NEIGHBORHOOD						
2:30	AROUND THE CLOCK						
3:00		30 Minutes With	Frontline N.Y.C.	Masterpiece Theater	Hollywood TV Theater	Firing Line	
3:30		French Chef	Lee Graham Presents				
4:00		Space Between Words	Film Odyssey	Jean Shepherd's America	Evening at Pops	Soul!	
4:30				Jazz Set			
5:00		Doin' It		Devout Young	Book Beat	Casper Citron	
5:30	SPECIAL OF THE WEEK	WNYC - TV NEWS					In and Out of Focus
		Can. 1/4 Hr	Scope	It. Panorama	Camera 15	New Horizons	
6:00		ELECTRIC COMPANY					French Chef
6:30		Around the Clock	Speaking Freely	Around the Clock	Return to Nursing	Around the Clock	Devout Young
7:00	Jean Shepherd's America	On the Job		On the Job	Around the Clock	On the Job	On the Job
7:30	One to One	N.Y. Report	Around the Clock	All About TV	Focus on the Arts	Brooklyn College	Firing Line
8:00	Magazines in Focus	Forsyte Saga	Frontline N.Y.C.			University Broadcast Lab	
8:30	30 Minutes With		Up South News	Lee Graham Presents	Casper Citron	Election/72	Evening at Pops
9:00	Brooklyn College	Forsyte Saga	The Police Commissioner	Film Odyssey	The Police Commissioner	Masterpiece Theater	
9:30	Focus on the Arts		In and Out of Focus		Consultation		Doin' It
10:00		Brooklyn College	Hollywood TV Theater		Soul!	Urban Challenge	All About TV
10:30	Urban Challenge	University Broadcast Lab				Your Right to Say It	
11:00							

Fig. 4—Schedule from 1972

federal funds for the development of new programs, and have met with substantial delays. One frustrating case has been that of the Housing Authority, which has long wanted to set up a training program for its engineers. After receiving an orientation from WNYC-TV and observing the Police Department in action, the Housing Authority applied for a federal grant. Two years later it received a grant to develop a program that would provide these men with certification and enable them to operate their equipment more efficiently. The Housing Authority engaged a producer to develop a training series, and his proposal remained under consideration for three more years. It still awaits final approval.

AUDIENCE PENETRATION

There has been no systematic assessment of audience numbers or types watching WNYC-TV. In a few cases, there does seem to be considerable interest. For example, Danbury, Connecticut and Mt. Kisco, New York are suburban towns with volunteer firemen who have had no formal training beyond an introductory orientation. The volunteer departments do have access to New York's professional Fire Department training on WNYC-TV. In fact, Danbury moved an antenna to the top of a hill so it could receive both the fire and police training programs from WNYC-TV.

As another example, the Department of Social Services used WNYC-TV in 1966-1967 to enable the Commissioner to communicate simultaneously with 47 welfare centers, representing a total of 23,000 employees. The Commissioner spoke on the latest developments in welfare administration, and then answered questions telephoned into the studio from the welfare centers. (Nineteen telephone lines were reserved for this purpose.)

Naturally, if funds were available, WNYC-TV would like to conduct a comprehensive audience survey. Although its audience is probably small compared with those of New York's VHF stations, it may be substantial in absolute numbers since the City has such a large population.

RELATIONSHIP TO NYC CABLE OPERATORS

A word should be said about the relationship between WNYC-TV and the two existing cable franchises in Manhattan held by TelePrompTer and Sterling Cable TV. These franchises went into effect in 1970, and partly because the origination of WNYC-TV predated them, they have never been integrated with WNYC-TV.

At the present time, due to the organization of the New York City government, the cable franchises and WNYC-TV are under two separate branches of government. The cable operators are responsible to the Board of Estimate, and WNYC-TV to the executive branch. This means that both cable operators have a dormant channel available for municipal programming and could, if they desired, actually commence such programming completely apart from WNYC-TV's activities. It also means that WNYC-TV receives no portion of the income from the cable franchises to support its operation.

In 1972, the Board of Estimate established an Office of Telecommunications to provide broader coordination for existing and future cable franchises in New York. This Office has not tried to change the existing situation thus far, though some type of merger of WNYC-TV activities and the city's cable operations is clearly possible in the future.

IMPLICATIONS FOR OTHER MUNICIPALITIES

WNYC-TV's experience holds several implications for the development of municipal channels on new urban cable systems. In spite of the technical differences between cable and over-the-air television, many of the organizational and programming characteristics will be similar, especially for municipalities operating a single channel. The following discussion examines five desirable characteristics for a municipal TV system: strong central guidance and facilities, well-trained personnel, resources to buy or produce attractive programs, highly motivated municipal service agencies, and continuing financial support.

Strong Central Guidance and Facilities

Because of limited budgets, WNYC-TV has been able to provide only a minimum of service to agencies wishing to go on the air. There is not sufficient staff, for instance, to advise the agencies on the best uses of television (media planning and direction), or to develop a coherent and attractive array of programs (a program management staff). Moreover, because the agencies have to use their own resources to produce their own programs, they purchase different types of cameras and recording equipment, not all of them compatible. In short, a strong central facility is needed to assure a well-balanced programming schedule, uniform standards of production, a well-trained technical staff, and the ability to take advantage of the latest technical developments in television production.

A city could therefore use its limited television resources much more effectively by establishing a strong central station that owned most of the equipment and provided technical advice to the agencies. A city just beginning to develop municipal television probably could establish such an operation easily, and it need not be elaborate. It should be noted that cities cannot count on using funds for this purpose that their local agencies receive from the federal government. The agencies receive federal assistance directly from their federal counterparts. A welfare agency, for instance, has a direct federal counterpart from which it receives funds for delivering social services. Some of these services may use television, but the funds would not, however, be directly accessible to a centralized municipal television station. The station itself has no real federal counterpart, and can turn to only a few sources of limited federal funds.

Well-trained Personnel

Municipal employment generally has a low turnover, and employee skills in areas of rapid technological change often become obsolete. Because television production has been such an area, it is necessary to provide continued training for employees. Most municipalities overlook the importance of this training, and may even overlook the necessity for formal indoctrination sessions at the beginning of employment. WNYC-TV, for instance, provides no formal orientation or training for its technicians, even though some are hired as trainees.

The success of a municipal TV system will depend heavily on the availability of qualified persons with up-to-date skills. With poorly trained personnel, the quality of production suffers and equipment may break down excessively. One partial solution for municipalities beginning television operations is to require equipment manufacturers to provide initial orientation and training as part of the purchase agreement, much like the service provided by computer companies and manufacturers of other office equipment.

Producing and Purchasing Attractive Programs

A well-balanced array of programs is necessary to attract audiences and provide municipal services that meet their needs. Program management includes consideration of such issues as program scheduling (e.g., they should be shown at hours when the target audience is available), the proper length and format of programs, and the mix of locally originated programs with those developed by professional associations, the federal government, and the public broadcasting network.

Attractive programs have not yet been produced for a number of important municipal services. Some, however, may be available through professional organizations, government agencies, and commercial sources. Municipalities must therefore plan on making sufficient funds available to purchase or rent these programs. Enough videotape should also be available so that municipalities can develop their own program libraries, especially for locally originated programs.

For program assessment it is of course necessary to determine audience numbers, types, and interests. This should be done routinely and continually. Though two-way interactive capabilities on the cable system may not be available, telephone surveys or the solicitation of postcards from the audience may serve the purpose.

Highly Motivated Municipal Service Agencies

It has been WNYC-TV's experience that most agencies are not predisposed to using television. Municipal stations must therefore constantly work to motivate them. Agency initiative is extremely important, since only the agency is likely to know the best ways of improving service delivery.

Most agencies become interested first in communicating with and training their own personnel. Police and fire departments are usually the readiest to take advantage of television; other agencies must be encouraged. The task becomes easier when agencies perceive opportunities to improve service delivery. The incentive for using cable will be especially strong if an agency finds that it can save or free manpower and other resources by using television. The municipal television station should try to identify such possibilities and point them out.

In general, it may be desirable to encourage each agency to have its own media and production staff. This staff would create and produce the agency's programs, with technical advice from the staff of the central studio facility.

Continuing Financial Support

Underlying all these conditions is the need for a sustained level of financial support for the entire television operation. A station subjected yearly to the whims of the budgetary process will have trouble attracting highly trained personnel and motivating them to produce high-quality work.

At present, the financial prospects for new operations like cable television may appear especially bleak. However, a successful cable franchise may eventually produce sufficient revenue to give some continuing support to the municipal channel. The greatest financial incentives, of course, will be created if the television programs can be shown to improve service delivery, and therefore to be an easily justifiable expenditure of public funds.

Part III

Uses in Education
Polly Carpenter-Huffman

I. INTRODUCTION

The past few years have seen the rapid growth of cable television throughout the United States, both in terms of the number of systems and in terms of the development of cable technology. With this growth has come an increasing awareness of the potential of cable TV for providing more than improved quality of signal in fringe areas for broadcast TV. Some people have gone so far as to predict a revolution in our way of life brought about by the many services that conceivably may be provided over cable, services that might greatly diminish needs for transportation. For example, one might "go to the office" or "go marketing" via cable.

But reality has been far different. As of March 13, 1971, only 18 percent of the operating cable TV systems in the United States originated programming locally that went beyond purely automatic services (for example, weather reports) or advertising.[1] Why? Federal Communications Commission (FCC) regulations, market uncertainties, and the high cost of providing local services have all impeded development.

Cable TV has the capacity to provide better, more, and new telecommunication services. Recognizing this potential, in 1972 the FCC issued a set of regulations for installing and operating cable TV systems.[2] These rulings have given rise to increased interest in the use of cable TV for public services.

PURPOSE AND PLAN OF THIS REPORT

One purpose of this report is to discuss the implications of the FCC regulations for those who will be concerned with the use of cable TV in education. But a more important objective is to stimulate the interest of the education community by acquainting them with the potential contributions of cable TV to education. This is the text of Chap. 2, "The Potential of Cable TV for Education." Our aim is to increase the involvement of educators in current cable TV planning and franchise negotia-

---

[1] *Television Factbook,* 1971-1972 edition.
[2] "Cable Television Report and Order," February 12, 1972.

tion activities to ensure that the contributions of cable are well exploited for education and not lost to commercial interests.

A basic stumbling block to the use of technology in education is the education community's lack of interest or ignorance. Working within institutions that define the process of education rigidly in terms of teachers, classes, and classrooms, educators have had little incentive to become involved in planning and implementing educational programs that make significant use of technology. And the materials and other features that are required to make the technology useful, such as instructional television programming of high quality, have gone by the board for lack of support. We discuss such difficulties in Chap. 3, "Barriers to the Use of Television in Education," both to help the reader steer clear of pits into which television has fallen in the past and to convince him of the necessity for his involvement and commitment.

This involvement should begin in the early stages of negotiations for award of the cable TV franchise. It is important that members of the education community play an active role in these proceedings, as many decisions made at this point can bear strongly on the future usefulness of the cable TV system to education. Possible roles for educators in the franchise negotiations and major franchise provisions they should consider are described in Chap. 4, "The Educator's Stake in the Cable TV Franchise."

WHAT IS CABLE TELEVISION?

Cable television is only one of several ways to distribute audiovisual information. Its essential features include a head-end[3] for amplifying and processing signals for cable transmission, coaxial cables to carry the signals to subscribers, terminal equipment connected to subscribers' TV receivers, and a group of antennas for receiving broadcast television signals for retransmission over the cable (see Fig. 1). Studios for local program production also can be connected to the head-end. Although the cable may be buried underground, it is usually strung from utility poles.

Other transmission media of particular importance to education are the familiar broadcast TV, closed-circuit TV (CCTV), and instructional television fixed service (ITFS). In the next Part,[4] we discuss the similarities and differences among these media and cable TV and point out situations in which one medium or the other would be more practical. At this point it will suffice to say that in many instances cable TV will offer multiple channels and privacy at a lower cost than the other transmission media (broadcast TV cannot, of course, offer multiple channels). Cable TV systems also have the potential to provide two-way service; most will not, however, for the next ten years.

[3] See the Glossary for definitions of technical terms.
[4] Part IV.

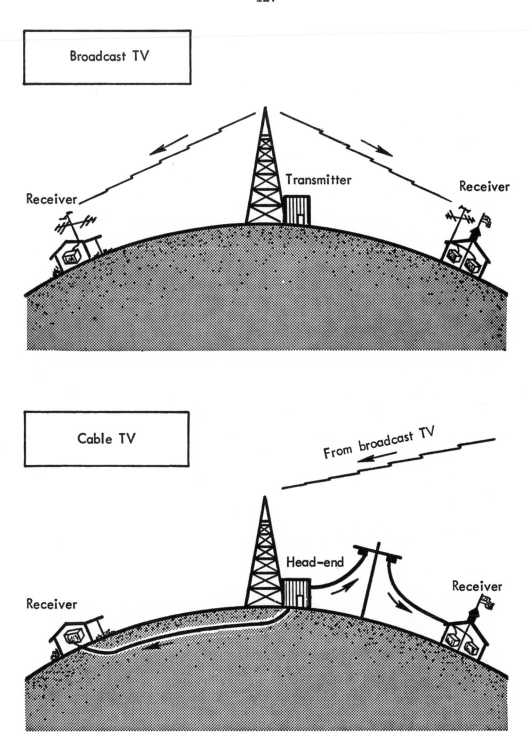

Fig. 1—A comparison of broadcast TV and cable TV

II. THE POTENTIAL OF CABLE TV FOR EDUCATION

We will first discuss general ways in which cable television can contribute to education, illustrating the discussion with whatever real-life applications exist and showing how the special features of cable could help television make even more effective contributions. Second, we will treat in detail the roles that cable television can play in instruction by describing the subject matter and teaching methods that are (and are not) suited to the medium.

The education community must consider cable TV as a means to attaining its goals—not as an end in itself. If a school system adds cable TV facilities to its inventory only to demonstrate its commitment to progressiveness or innovation, chances are very high that all that will be accomplished is a rise in the cost of running the school; the cable TV system will be just another dispensable frill. To ensure that the cable system is being used as a means to desirable ends, the contribution of cable TV should be *essential to the school program* and *not performed as well—if at all—in some other way.*

GENERAL CONTRIBUTIONS

We shall discuss four significant uses of cable TV in education:

- Making education physically more accessible to students.
- Providing educational services that cannot readily be provided by other means.
- Improving the quality of education.
- Decreasing the unit cost of education.

Examples of each use will be drawn from recent experiences with educational and instructional television (ETV and ITV). ETV is a broad term that encompasses all types of educational programming, including information, instruction, culture, public affairs, and entertainment. ITV, on the other hand, refers to educational efforts using television "which have as their purposes the production, origination, and distribution of instructional content for people to learn.... Instructional television

128

is closely related to the work of organized formal educational institutions."[1] Public television (PTV) is that portion of ETV that is directed to the general community.[2]

ITV may use any expedient combination of broadcast TV, ITFS, closed-circuit TV, or cable TV; PTV is usually carried over broadcast or cable TV. Because cable TV has appeared relatively recently in most communities, however, most of the examples of its potential use will be drawn from systems using some combination of the other transmission media. This does not make these examples any less relevant in considering the use of cable TV, except for systems covering a large geographic area. In that case, cable TV might not be available throughout the area of interest, and other transmission media might be needed in addition to cable TV.

Whether it is a practical medium will depend on local circumstances. Of course, few examples demonstrate the instructional potential of the multichannel and two-way applications made possible by cable TV. In these instances, hypothetical examples must be provided.

Improved Accessibility

A cable TV system can provide links between schools and homes. One apparent advantage, on first thought, is that students who must be absent because of illness or handicap can join their classmates at school via cable. In fact, this has been done in several systems. In Overland Park, Kansas, a student confined to his home after an operation took a course in American history over a special cable setup. He responded to his teacher's questions by using a keyboard attached to the set, and she was able to see him through a camera at his bedside.[3]

Consider, however, what would be required if the cable system were to be put to such a use on a broad basis. Assume that the average number of classrooms in a school is 22, the average number of students in a classroom is 25, and the average absence rate is 10 percent. Assume further that each teacher is free to conduct his classes as he thinks best, within certain broad guidelines. This means that the work of each class is different. If only one channel is to be used from the school to the homes of its students, an average of less than three students (10 percent of 25) will be able to keep up with their class via cable. (Obviously, on the average, 22 channels would be needed to provide the service to all absent students.) This is clearly an uneconomic use of the cable TV resource.

A less costly application would be to provide programs of general enrichment or review for homebound students. These could be designed for the average student at a given grade level and could be accompanied by assignments for student work that would be checked periodically or upon the student's return to class.

It has also been suggested that review and remedial materials could be presented to home viewers after class hours. Unless ITV becomes much more attractive to students, this seems unlikely to be successful below the college level; few students will be motivated strongly enough to continue schoolwork after the regular school day without a teacher's supervision. With the exception of the dedicated student, home viewers must be enticed into viewing by captivating or engrossing program-

[1] *Educational Product Report, No. 31,* p. 16.

[2] Carnegie Commission on Educational Television, *Public Television: A Program for Action.*

[3] Price and Wicklein, *Cable Television: A Guide for Citizen Action.*

ming. In addition, it may not be desirable to extend the school day for the elementary or secondary school student.

Of course, the school system may already be geared to the use of television for instruction to the extent that all students in a given grade and subject view the same program at the same time. This is the case in the Hagerstown system,[4] which will be described below. In this event, the total number of channels needed for at-home instruction would be the same as the number used in the system. In Hagerstown, six channels are used to serve grades 1 through 12.

The large-area coverage of broadcast TV soon encouraged its use for students who found it difficult to attend school regularly. Some of these students, such as women with small children, are homebound; others are working during the normal school day and are too far from technical schools or other institutions to attend night classes. There is also convincing evidence that large numbers of adults would make use of television for continuing education whether or not these programs led to course credits in a formal institution.

An outstanding example is the Chicago TV College which, since it was established in 1956, had provided courses for credit to over 98,000 students by the fall of 1970. Data indicate that "there are about 250,000 regular viewers each semester, and perhaps 500,000 frequent, but not daily, viewers of each semester's courses."[5]

Several networks have been established to link graduate students at geographically dispersed points, such as colleges and industrial firms, with graduate schools at institutions of higher education. One of these is the TAGER[6] network in Texas, which links program-originating facilities at Southern Methodist University with 9 institutions of higher education and 11 industrial plants in the Dallas-Fort Worth region (see Fig. 2). Similar systems are now operating in California and Oklahoma. Despite the relatively small number of students who participate in such systems (approximately 2000 students per semester in the Texas system),[7] the high cost per student is justified because conventional graduate education would cost at least this much even if the students had ready access to formal classrooms.

A *single* cable TV system would be inappropriate as a transmission medium for these networks because of the relatively large geographic areas they cover. According to Feldman,[8] a single CATV system that covers an area of much more than 5 to 6 miles in radius from the head-end suffers signal degradation at the outer periphery. Cable TV could, however, be used for local distribution within such a system. For example, the cable TV systems being planned in 1971 for Dallas and Forth Worth[9] could be used to distribute the Southern Methodist programs to residents as well as to industries in those cities. Alternatively, videotapes and films produced by Southern Methodist could be distributed by mail to the program-originating facilities of these cable TV systems, saving channel space on the larger network for the graduate programs with small audiences in each city. Of course, if distances are small or if sufficient subscribers exist to support several cable TV

---

[4] Wade, "Hagerstown: A Pioneer in Closed-Circuit Televised Instruction."
[5] Bretz, *Three Models for Home-Based Instructional Systems Using Television.*
[6] The Association for Graduate Education and Research.
[7] TAGER, *Annual Report, 1969-70.*
[8] "System Designs for the Dayton Metropolitan Areas."
[9] *Television Factbook,* 1971-1972 edition.

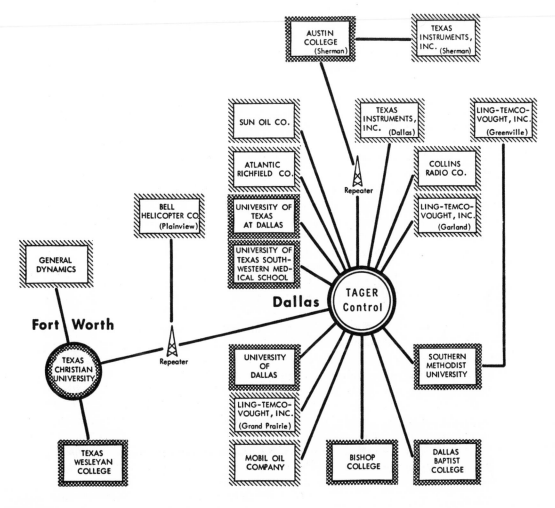

Institutions of higher education

Industrial companies

The circles symbolize the omnidirectional microwave transmission which is available in Dallas and Fort Worth. All other transmission is via microwave beams.

Fig. 2—The TAGER television network

systems linked together, as proposed for Dayton, Ohio,[10] cable TV could be used as the main transmission medium.

England's Open University and Germany's Telekolleg are other examples of successful ventures in using TV (along with other offerings) to make higher education more accessible to people who find it difficult to attend established institutions. Unlike the preceding examples, however, both of the European ventures were undertaken wholly independently of established schools, although each draws heavily on talent within the existing academic structure and supplies credits to students who complete the course work successfully and pass the examinations.

Despite the ubiquitous U.S. system of elementary and secondary education that brings schooling to nearly all below the age of 18, there are probably a large number of older people who find it difficult to attend regular schools and who would take advantage of more convenient opportunities to further their education. The responses to Telekolleg and the Open University in Europe, and the Chicago TV College and Sunrise Semester in New York, all suggest that there are many adults who would participate in TV-based programs in secondary and higher education. The 1970 census, for example, reports that nearly 48 percent of the population over 25 have not completed high school. In 30 selected "Standard Metropolitan Statistical Areas," over 55 percent of the nonwhite population do not have a high school diploma.[11]

Also, it is estimated that 20 million adults are functionally illiterate,[12] many because their native tongue is not English. These people either do not have a convenient school in which to learn English or are cowed by formal educational institutions. But there are enough of these people concentrated in large cities that TV programs for them would be economically feasible. For example, of the 1.5 million persons in low-income areas in New York City, 331,000 are foreign-born, most of them of Puerto Rican origin. (In the same areas, more than 65 percent of the population have not completed high school.)[13]

Unfortunately, much of the audience with the greatest need for this type of programming will not be on a cable system, as people in the lower income levels usually can't afford the monthly fee for cable service. However, it is possible to provide the target audience with cable service as part of the project. An intermediate solution is to provide neighborhood viewing centers connected to the cable system.[14] A concomitant advantage of this solution is that it supplies the audience with the social stimulus of its peers, along with the televised programming.

A crippling problem with many programs for adult education is that they are consigned to the very early morning hours. Sunrise Semester's name aptly describes the situation. Although many people may be sufficiently motivated to view educational programs at these hours, it is likely that more convenient viewing times would greatly increase the audience. By alleviating present restrictions on educational channels, cable TV could reach this large audience.

As suggested earlier, there are undoubtedly large numbers of adults who would watch educational programs that are not for credit. Despite the complications of

[10] Feldman, "System Designs."
[11] *Educational Attainment in 30 Selected Standard Metropolitan Statistical Areas: 1970.*
[12] "Interview with Dr. Donald G. Emery, Director National Reading Center," *Reading Newsreport.*
[13] *Employment Profiles of Selected Low-Income Areas.*
[14] Baer, op. cit.

attending class at night, the University of California at Los Angeles Extension typically enrolls many students not seeking formal credit in extension courses. For example, over half of the students in an extension course in anthropology in the winter of 1971 were not for credit. And, despite its uncomfortable viewing hour, Sunrise Semester lures an estimated 1.25 million students from their beds; most of these students do not take the courses for credit.[15]

People who take advantage of such learning experiences have passed the age at which they feel the need to participate in educational rites of passage; they are involved with learning for the contributions it can make to their enjoyment of life and self-development. They may be older people whose primary careers have ended, women whose children have left home, adults seeking to brush up on salable skills, or adults wishing to participate in intellectual, political, and cultural activities. The potential population of such people is very large, numbering in the tens of millions. Traditional schools, however, are youth- and degree-oriented, and few schools have attacked the tasks of providing worthwhile educational experiences for the older population. The challenge will be to devise educational programs that will have a general appeal to these people. On the other hand, removal of requirements for accreditation will allow producers of these programs the freedom to take advantage of the television medium to devise programs that are both captivating and intellectually stimulating.

We have been discussing audiences of adults whose primary educational needs are shared by large numbers of people—people wanting high school equivalency courses, literacy training, or courses of general cultural, social, or economic interest. Because broadcast TV must appeal to large audiences to be economically attractive, it has been limited in its ability to serve specific needs for vocational or professional education. Only where the cost of conventional instruction is high, as in graduate engineering education, medical education, or military training, has television played an important role. The multichannel capacity of cable, however, opens the possibility that it might help in meeting needs for specialized vocational and professional education in various situations.

It might be argued that existing institutions for vocational and professional education can take care of these needs, that anyone really wanting such an education will find a way to get it from the current offerings. Two facts suggest that this may be a limited view. Foremost is the lack of people with middle skills (skills requiring some education or training but not formal degrees). For example, in the Akron area, there are chronic shortages of auto body repairmen, draftsmen, registered nurses, production machine operators, and tool and die makers.[16] Second is the expressed need for additional education or training in certain segments of the population. Of the 1.2 million adults in low-income areas in New York City, over 12 percent stated that they were hampered in working by lack of skill, experience, or education. Two-thirds of these are employed, suggesting that work experience may sharpen a person's awareness of education and training needs. Currently, the Office of Education, in one of its programs in Career Education, is experimenting with the possibility that television and other media can be used to foster such an awareness

[15] Efron, "Peter Chelkowski, Ph.D., I Love You."

[16] Ohio Bureau of Employment Services, *Occupational Survey, 1970-71, Akron Metropolitan Area.* The total number of positions in the middle-skill areas was around 20,000.

in the homebound population.[17] Beginning in the fall of 1972 in Providence, Rhode Island, a concerted campaign was conducted through the mass media to bring the unemployed and misemployed to career counseling services offered by the Educational Development Center.

Added Services

Probably the best known examples of the ability of television to add significantly to the education provided by the schools are Sesame Street and the Electric Company. Sesame Street is viewed by surprisingly large numbers of children of preschool age and in the first grade. In 1970, the number of viewers ranged from a low of 32 percent of the preschoolers in Washington, D.C., to a high of 100 percent in Boston. A Nielsen survey from November 1969 to March 1970 estimated the average audience for Sesame Street at 1.1 to 2.4 million.[18] By appealing to such a large audience, the producers of Sesame Street have been able to bring large resources to bear on a pervasive educational problem—the teaching of basic skills. The $600,000 that went into research and the $24,000 per hour for production[19] resulted in programs with high student (and adult) appeal and teaching effectiveness. Some schools now form two classes for students entering the first grade—one for students who have watched Sesame Street and one for those who have not. Similarly, the Electric Company is making strides to solve the problems of the nonreader in schools, where the teaching of reading nominally stops above the lower primary grades.

Less well known than Sesame Street is the Home Oriented Preschool Education (HOPE) program of the Appalachia Educational Laboratory in Charleston, West Virginia. HOPE combines daily television lessons, weekly home visits by paraprofessionals, and weekly group sessions to educate the otherwise isolated preschooler in rural areas of Appalachia. Children and parents are reported to participate actively in the program; the National Center for Education Communications has selected HOPE as one of ten most innovative programs of recent years.[20] If successful, HOPE may provide a model for preschool education for which cable TV is particularly appropriate.

By far, the bulk of ITV is devoted to supplementing the regular curriculum with materials that would be difficult for the classroom teacher to acquire or devise. Because of the supplementary nature of these programs, their use is left to the discretion of the classroom teacher, and often little or no preparatory or follow-up activity is prescribed. This probably explains the low usage of ITV; in 1970, 4590 hours[21] (at most) of ITV were broadcast during a week in March *throughout the whole United States.* Other figures suggest that 12,000 school districts could have used ITV broadcasts in 1970; a utilization rate of only one hour per district per week would more than double the amount of ITV programming provided. Of course,

---

[17] *Career Education Research and Development Program.*

[18] Samuels, *The First Year of Sesame Street.*

[19] Connell, "How Sesame Street Raised and Spent Eight Million Dollars."

[20] "Experts Praise AEL's Home-Oriented Preschool Program."

[21] This figure was derived from data provided in Rockman, *One Week of Educational Television.* I have adjusted the data upward to account for stations and school-owned systems that failed to reply to the questionnaire.

several school districts may use the same program simultaneously; this practice is not common, however. The low utilization rate is not surprising when we note that of the schools having TV sets, 52 percent have less than one for every five classrooms.[22]

In some states, statewide networks have been set up to provide television to the schools. One of these is KET, Kentucky Educational Television, serving the length and breadth of Kentucky from Lexington. The network is laid out in five divisions, comprising 13 educational TV stations linked by closed and broadcast circuits. Programs reach 325,000 students in 175 of Kentucky's 195 school districts. All five divisions currently carry the same programming, but it is planned to tailor programming to interests in each region when program material becomes available.

The objectives of KET are to

(1) Improve education in Kentucky classrooms; (2) make teachers more effective without working them harder; (3) reach every schoolchild in the state; (4) enrich the community around each school; (5) supplement college and university instruction; and (6) help state agencies serve Kentuckians.[23]

In spite of the supplementary nature of most uses of television in the schools, some of these applications have been successful, when measured by teacher demand. An example was the Ottawa cable TV system for televising films and videotapes chosen by the classroom teachers. The system used 12 channels to service 130 learning areas in two primary schools, one junior high school and one high school. The schools were connected to a library of 2600 videotapes and films, selected by a committee of teachers.

The typical terminal had a TV receiver and a telephone with a direct link to the library. Each receiving location was also equipped with a cross-referenced catalog of the contents of the library, to inform users of its contents. Having decided what program he wanted to view, the user called the library and gave his program request to the librarian. The librarian allocated the channel on which the program would be shown and gave a time of transmission. Transmission normally began within 60 seconds of the receipt of a request.[24] In this system, the television receiver played the role of a movie projector or videotape player in the classroom, thereby alleviating problems of logistics and equipment unreliability that usually deter teachers from greater use of films and videotapes.

However, in an evaluation of the Ottawa system, the Ontario Institute for Studies in Education determined that such a system

is unlikely to do much more than double the average student's exposure to films and television. Furthermore, [it] will be used predominantly for enrichment.

Hence,

It does not seem worthwhile to add at least 2 percent to a school board's annual budget until the expenditure can result in some significant impact

---

[22] See the Appendix.
[23] Belt, "Education in Kentucky—By Television."
[24] Billowes, "On-Demand Educational Television Program Retrieval System for Schools."

on the teaching and learning tasks to which schools are particularly suited.[25]

The public schools of Washington County in Maryland have been using a closed-circuit TV system as an integral part of instruction since 1956. This system (usually referred to as the Hagerstown system)

> has made it possible for pupils: (a) to take, instead of extremely limited amounts of elementary-school science and one-year science courses in high school, a sequence of science courses that extends from grades 1 to 12; (b) to have special television instruction by experts in art and music; (c) to take a modern-language course beginning in grade 3; and to take an advanced mathematics course in high school that covers calculus and analytics.[26]

Because special skills are called for, these services would have been too expensive to be provided by classroom teachers. Thus, the Washington County students had a much richer educational experience than they would have had without television. The Hagerstown experience might well point the direction to maintaining a diversity of educational offerings in view of school budgets whose buying power continues to shrink.

Most teachers are constantly involved not only in teaching others, but in learning themselves. Salary schedules for teachers are typically geared to years of teaching experience as well as to additional education intended to contribute to teaching competence. In addition, school systems usually provide *in-service training,* training aimed at improving teaching competence in special areas or in developing skills to be used in innovative programs. So it is surprising that only small percentages of ITV programming are being devoted to education as a subject area. For 1970, Rockman gives between 1 and 5 percent of ITV programming to education as a subject area.[27]

It would seem that the hours after school or in the early evening would be especially suitable for providing in-service training and education courses for credit. Schools already making use of television are in an excellent position to supply this service to their teachers. In fact, Oregon State University, which has an extensive CCTV system, provides the following services to the School of Education:

- Nearly all the Methods classes in the School of Education use the TV Center for micro-teaching[28] purposes by teacher-trainees utilizing a variety of formats ranging from the teaching of short lesson segments to full class hour presentations.
- The School of Education Counseling Center in Education Hall has three TV-equipped rooms linked to the Kidder Hall TV studio by private line. At any time, a counseling session may be recorded on one of the studio machines in complete privacy and transmitted back to the counselor for evaluation and later erasure.[29]

---

[25] McLaughlin et al., *Educational Television on Demand.* Although 2 percent seems small, remember that most school systems are already bumping against their budget ceilings and that this sum must come from the already very small amount (from 10 to 15 percent) allocated in a typical school budget for all items other than salaries.

[26] Wade, "Hagerstown."

[27] *One Week of Educational Television.*

[28] Micro-teaching is practice teaching of small classes. Usually, the trainee is videotaped during the session for later self-evaluation and critique.

[29] Johnson, *Cable Television and Higher Education.*

In addition, many services are provided to the School of Education along with other University departments, such as transmission of course review materials before final exams to residence hall viewing rooms and, when desired, to over 4000 homes and apartments.

Two other instructional uses should be mentioned. One is the obvious use of television to train students in techniques of production and in operation of television equipment. Although professional preparation will be most common in senior high schools and higher education, students at all levels will find some participation in the television medium stimulating. For example, in Newburgh, New York, young people developed their own production facility in which they produced videotaped interviews, discussions, documentaries, and reports for the cable system. The project was supported by The Ford Foundation.[30]

Television can also be used for giving examinations, particularly those that test students' abilities for audiovisual recognition and discrimination. The national drivers' tests are an obvious example. TV-administered testing can also help to ensure uniformity of testing conditions from classroom to classroom so that student scores will be more comparable. Trained testers can give directions and administer parts of the test requiring oral stimuli, a stumbling block in many teacher-administered tests.

A few school systems have recognized that cable TV offers an excellent opportunity to involve the community more closely with the schools:

> Educators have come to recognize that *informed* parents and taxpayers are likely to be more understanding and supportive of the schools' efforts, and that parent involvement during (and even as a forerunner to) the formal teaching process is highly desirable.[31]

Thus, the Willingboro Township Public Schools in Willingboro, New Jersey, has instituted a wide variety of school-to-home programs over the public cable system. Among them are

- Coaches Corner, a discussion of high school athletics.
- Transmission of student-produced videotapes on such subjects as "How Congress Works" (made in Washington, D.C.), a student interview with the assistant administrator for the Federal Aviation Agency on a proposed jetport in the vicinity of the school district, another student interview with a state senator on the Anti-Ballistic Missile System.
- Taped sessions of the Social Action Committee established at the junior high schools to work for better human relations.
- Entertainment by student musicians and dancers.
- Information about new school programs.
- Instruction in reading and arithmetic for preschoolers, accompanied by manuals for parent participation.

Improved Quality

It could be argued that adding services as described above improves the quality

---

[30] Price and Wicklein, *Cable Television.*
[31] Reuben, "Using Cable Television To Involve Parents."

of education. Here, however, we are addressing uses that improve quality by substituting for some of the normal activities of the classroom teacher. Although such usage is far less common than supplementary uses, some school systems in the United States have been making routine use of television for classroom teaching for more than ten years. One of these is Hagerstown, mentioned above. Basic instruction in both required and elective subjects is given by television in the Washington County schools served by the studios in Hagerstown. Nearly 30 programs a day are transmitted over up to five channels in the closed-circuit system. Subjects taught include science, music, arithmetic, art, language, math, English, social studies, biology, history, reading, algebra, physics, and advanced math; grades range from 1 through 12.

When the Hagerstown system was first implemented, there was considerable concern about the effectiveness of TV as a teaching tool. Extensive studies were conducted with the support of the Fund for the Advancement of Education and The Ford Foundation.[32] Achievement data on Hagerstown students indicated that they learned more, by and large, from television than they would have learned in the regular classroom. As is frequently the case when several subjects and grade levels are involved, however, results varied by subject and grade. Achievement in arithmetic and upper-level mathematics, as measured by the Iowa Test of Basic Skills, was greatly improved after television had been introduced.[33] Similarly, there was a notable, sometimes spectacular improvement in science, as shown in Table 1. Note that television seemed to be most helpful to students of lower ability. On the other hand, achievement in reading and English was not as impressive, and small gains were recorded in social studies and language. The gains were explained "in large part, at least, as a result of the new curriculum made possible by television."[34]

For over ten years the Psychology Department of the University of Akron has used closed-circuit television to teach general psychology to undergraduates. During this period, the University has conducted extensive research on the effects of televised instruction on student learning, attitudes, and other matters. In 1972, Dambrot summarized this experience by stating that teaching by television

> is a viable educational technique that need not dehumanize or mechanize the learning process. With proper presentation and administration, a televised course has the capability of transmitting a quality level course to a large mass of students.
> Television at the university level seems to be preferable to the practice of inexperienced graduate teaching assistants teaching freshman and sophomore classes.[35]

Decreased Unit Cost

For many years, the quality of education has been measured by the cost of the resources used—teachers, buildings, and materials—rather than by the amount of learning produced or the number of graduates who get jobs or enter college. As a

[32] Wade, "Hagerstown."

[33] The comparisons were made among students at the same grade level from year to year.

[34] Wade, "Hagerstown."

[35] "General Psychology over Closed-Circuit Television."

Table 1

COMPARISON OF STUDENT GROWTH IN SCIENCE ACHIEVEMENT

Students Taught Conventionally		Students Taught by Television	
Average IQ	Growth (in months)	Average IQ	Growth (in months)
117	12	118	15
100	11	100	14
83	6	83	13

result, it has been increasingly expensive to teach a student a given subject—and apparently without a commensurate increase in the amount and quality of learning. Measuring quality of product by resource use encourages educators to put more resources into schooling; practically no one has advocated economy because of the fear that a cutback will somehow damage the students.

Thus, it is not surprising that technology has been *added* to other elements of instruction, rather than being used to *replace* them. In fact, today's byword is "You can't expect to save money in education by using technology." Until technology is used to replace other elements of instruction for some tasks, particularly those performed by the classroom teacher, the heavy burden of teacher salaries will continue to grow and schools will not realize the savings that are potentially available.

The Anaheim system is one of the few examples of the successful use of TV to decrease the unit cost of education as well as to improve educational quality. The Anaheim scheme includes the redeployment of fifth- and sixth-grade students into rooms that can accommodate 75 students during the televised instruction under the direction of two teachers. This plan saves two classrooms and two teachers for every 150 fifth- and sixth-graders. The result is that in 1959-1960 the TV instruction handled 12 percent of the curriculum at less than 5 percent of the total annual district budget.[36]

ROLES IN INSTRUCTION

Let's now look specifically at the uses of television, particularly cable television, in instruction. For what subjects is television an appropriate or inappropriate medium? For what teaching methods is it especially suited or unsuited?

[36] Bretz, "Closed-Circuit ITV Logistics."

Subject Matter

Television has been shown to be very useful in teaching facts, concepts, and cognitive skills, that is, objectives falling into the cognitive domain in Bloom's taxonomy.[37] It is also useful in demonstrating motor and psychomotor skills and, to the extent that demonstration is sufficient for instruction, television can carry the burden of instruction even in some performance areas. Calisthenics is an obvious example. In fact, many demonstrations may actually be made more effectively via television than in person because the visibility of small or hidden movements may be improved.

Television must be supplemented by other instructional resources if

- Student performance must be monitored for safety reasons (learning to put out fires).
- Student performance must be monitored at each step so that successful progress may be made toward an end result (building a complex piece of equipment).
- The student must work with special materials or equipment that he would not normally have access to (aircraft repair).
- The student must learn an interactive skill, that is, one that requires him to work with others (team sports).

Most subjects with a large content of instruction inappropriate for television will fall in vocational areas (building trades, repair of equipment or machinery, service occupations) or in areas requiring large amounts of student performance, especially if interactive skills are required. In addition to team sports, examples of the latter are coordinated operation of a piece of equipment such as crewing a sailboat, group performance for an audience such as drama workshops or orchestra, or less rigidly prescribed group interactions such as public speaking or debating.

It is not doing the television medium justice, however, to consider its appropriateness only in terms of activities for which it can *substitute* for classroom instruction. First, a wide variety of audiovisual experiences can be provided the student *only* through the media of television or film, and, second, many others would be so costly to supply by direct experience that they are not feasible. In the first category are experiences that would endanger life or property, such as demonstrations of the spread of forest fires or of the hazards of braking an auto suddenly on an icy street. Also in this category are one-of-a-kind events such as the Apollo moon shots or Presidential addresses. And, of course, film or videotape is ideal to capture time-lapse or slow-motion photography, animation, pictures that are larger- or smaller-than-life, and sequences that depend for their effects on a mélange of images—all of which serve to expand the students' perceptual and conceptual capabilities manyfold.

In the category of presentations that can be provided more practically via the film or tape medium are those that involve

- Equipment that is either too expensive, unwieldy, or fragile to supply in the classroom, such as an electron microscope.

[37] *Taxonomy of Educational Objectives.*

- Facilities or locations that are difficult for the students to visit, such as Jacques Cousteau's underseas laboratory.
- Prominent people or people with unique skills to whom access is limited.

Some examples of these kinds of uses have already been cited on p. 13.

And, finally, because more resources can be expended on the production of film or television programs than the classroom teacher usually has at his disposal, such programs can be of better quality and without doubt more effective in bringing about student learning.

Television's unique capacities for presenting subjects that are inaccessible to the classroom teacher have been tapped in the field of instructional television for more than a decade. Table 2, which gives the percentage of ITV programming hours devoted in the United States to various subject areas during a week in 1970,[38] is suggestive. The data reveal that classroom teachers find ITV most useful in supplementing[39] courses that are not skill-oriented and that draw from a large body of nonclassroom experience (such as the first two subject-area categories) or that require skills the ordinary teacher may well lack (such as music and art). The data on language arts and literature are the apparent exception; quite possibly most of the programs in these areas are also not skill-oriented. Book reviews and dramatic productions are favorite topics for such television programs.

Table 2

ITV PROGRAMMING BY SUBJECT AREA

Subject Area	Percentage
Social and behavioral sciences	28
Physical sciences	19
Language arts and literature	16
Music and art	14
Mathematics	6
Foreign languages	5
Health, safety, and physical education	4
Other	8

Because of the traditional stress in education on the cognitive and psychomotor areas of personality development, little is known about the use of television in bringing about affective learning. What is considered, if anything, are the students' (or teachers') attitudes toward the use of television for teaching.[40] Even in the older

[38] Rockman, *One Week of Educational Television.* These data refer to ITV only. Sesame Street is not classed as ITV because it is not intended for in-school use, but during the survey week, it accounted for nearly 20 percent of all ETV programming that was not ITV.

[39] Recall that almost all uses of ITV are for supplementing basic instruction.

[40] Chu and Schramm, *Learning from Television.*

medium of film, "Much thought has been given but little experimentation has been done in this difficult area."[41]

The advertising industry has exploited the ability of television to mold public taste. Evidence indicates that television may bring about superficial shifts in preferences (say, from one brand of cigarettes to another), but major shifts in attitudes (say, from smoking cigarettes to not smoking) are much harder to bring about. "In fact, the mass media in general have proved to be ineffectual as tools for profoundly converting people. Studies have shown that persons are more likely to heed spouses, relatives, friends, and 'opinion leaders' than broadcasted or printed words when it comes to deep concerns."[42]

The implication of this quote is not, of course, that films and television do not affect people's attitudes, but that to affect fundamental attitudes the media must be reinforced by people whose leadership is accepted by the viewer. Cable TV can possibly be a powerful tool in this regard because it will facilitate the participation in the television programming of local leaders, perhaps personally known to many of the students. How effective such a tactic would be in bringing about attitude change must be left for future experimental test.

Teaching Method

With much ITV programming, the classroom teacher is transferred to the TV screen. He uses either straight lecture or a combination of lecture and demonstration. In the hands of accomplished teachers, these techniques can be captivating, as Continental Classroom, a forerunner of Sunrise Semester, proved. They are ineffective, however, in the hands of the average or even better-than-average teacher, especially when directed to audiences accustomed to the talent and expertise of commercial television. Conventional techniques also allow for little or no active viewer involvement and little or no adaptation of instruction to the needs and interests of individual students or groups.

The charge that television is a *mass* medium for *passive* viewing is damaging to its use in education, particularly now, when there is heavy stress on tailoring instruction to the needs of the individual student and on the active involvement of the student during instruction. Meeting the needs for individualization will be far more difficult than meeting the needs for involvement, however. This will be illustrated in the following discussion with regard to both broadcast TV and cable TV.

Methods for the Mass Media. In the early days, some attempts to involve the TV student consisted of showing televised instruction being given to a class of students, who asked questions and discussed the material on TV during the program transmission. The TV student supposedly thought of himself as part of the class. It soon became evident, however, that the viewers did not respond well to this technique, because the loss of eye contact with the TV teacher made them even further removed from him.

Despite its lack of requirements for overt response, Sesame Street obviously involves the viewer. This is accomplished primarily by amusing him, by piqueing his curiosity, and by presenting a continuous variety of attention-getting stimuli.

---

41 C. R. Carpenter, "Instructional Film Research."
42 Etzioni, "Human Beings Are Not Very Easy To Change After All."

Sesame Street's success in this regard was largely a result of steps taken during the development of the series to capitalize on the preferences of the target audience for existing film and television materials. The researchers devised

> an experimental method ... to measure a child's interest in a given piece of material by continuously recording his visual orientation toward or away from the television screen during the presentation.
>
> [This resulted] ... in an index of the relative appeal of a broad range of entertainment and instructional films and television programs. Furthermore, graphing the fluctuations in audience interest in a particular program permitted the researchers and producers to analyze the program from moment to moment to discover those elements which were most compelling of attention....[43]

The result was a series of one-hour shows that held many children spellbound before the TV set, disproving the traditional notions of the short attention span of young children. To say that the child who is watching Sesame Street is passive is to miss the overriding importance of covert (i.e., unobservable) response. In fact, all but reflex responses are initially covert responses.

Linear programmed instruction is a method of getting the student actively and continuously involved with the instruction. In such instruction, concepts and facts are presented in small increments or steps; after each step, the student is asked a question to test his understanding of the fact or concept. After a suitable pause, during which the student is expected to respond, the program usually provides the right answer so that he may check himself.

Linear programmed instruction may be used by either broadcast or cable television. It doesn't require that the student have a direct effect on the program. Its primary teaching advantage is that it encourages (or can be made to require) the student's active participation in the program. A disadvantage, however, is that the pace of presentation cannot be adjusted to the rate with which each student grasps the materials. If students have highly disparate learning rates, the program will be either too slow for some students, too fast for others, or both. This problem may be largely overcome by separating students of similar learning rates into groups and pacing the presentation appropriately for each group. Perhaps as few as three groupings would be sufficient to accommodate the spreads of learning rates that would normally be encountered. Note, however, that this solution automatically multiplies by three the amount of program transmission required and makes a multichannel system more attractive than one using only one channel.

In one system developed in Pennsylvania, International Correspondence Schools and TuTorTape Laboratories developed a response device attached to the television receiver that controlled the portion of the screen to be shown.[44] When a question was asked, a possible answer was given in each of the four quadrants of the receiver. The student pressed the button corresponding to the response he felt was correct. If he was correct, his answer was repeated to reinforce the concept. If incorrect, he was told in the numbered square activated by his button why he was wrong and then given the correct answer and an explanation. Although this system adapted the program to the student in a rather limited way, it could be an effective

[43] Kratochvil, *Sesame Street.*

[44] Sivatko, "Newest Teaching Method Provided by Educasting."

teaching tool for the slow learner. The technique suffered from the disadvantages that each student needed his own TV monitor, which increased the cost of an in-school system, and that the degree of individualization provided seemed too limited to justify the cost.

Methods for Dedicated Systems. We noted earlier that the uses of cable TV would be similar, by and large, to the uses of other media of TV transmission in education. This is true in the broad sense; that is, cable TV can serve the same general goals as can any TV transmission medium. There are significant differences, however, stemming from the relatively large number of channels that cable TV may make available at relatively low cost to the schools or other educational endeavors. This will greatly alleviate scheduling difficulties and may serve a greater diversity of needs and interests. The privacy of the dedicated system contributes to this possibility as well.

Since a single channel can be directed to a small audience, its transmissions can be made more responsive to the viewers. The Ottawa system described earlier, in which teachers received on demand films and other audiovisual material on the classroom TV sets, is an example of the responsiveness that can be obtained in a multichannel system. Also, the university-level ITFS systems using talk-back, mentioned previously, are examples of what can be done in the way of on-line feedback with live programming. In these systems, the lecture is transmitted live to receiving points scattered over a large geographical area. Students gather in classrooms at each receiving point and can transmit questions to the lecturer in the studio via either a special audio channel integral to the system or a telephone circuit. Thus, instructional programming is further tailored to the needs and interests of the students on the basis of their responses during program transmission.

The lecturer is not overloaded with questions because only one transmission can be sent from a classroom at a time. Users report that this system encourages productive small group discussions within the classroom and that the single talk-back channel is more than sufficient to carry questions that cannot be resolved by classroom discussion. Many courses at the graduate level exist primarily to provide students with direct contact with leading scholars; hence talk-back systems are especially useful at this level.

It will be noted that a talk-back capability can readily be obtained by renting telephone lines. Thus, the eventual two-way feature of cable TV is not essential to such a use, although in most instances telephone company charges will exceed the cost of providing the talk-back capability within the transmission system.[45]

Another direct feedback technique could be used to provide a running record and a final summary of student responses during a televised presentation using programmed instruction. One such system is EDEX, which has been in use for some years in schools, industry, and the military. The instructional material for EDEX is prepared in a linear programmed format, with multiple-choice questions and pauses for student response. The response system is a push-button device attached via the cable to a simple computer at the origination point. This computer displays the responses of each student and the number of right responses for each question

[45] For example, Martin-Vegue et al. estimate that telephone lines would cost nearly twice as much over a ten-year period as the cost of providing the same capability via ITFS response stations. The cost of using time-shared talk-back would be less than a fifth that of using telephone lines. "Technical and Economic Factors in University Instructional Television Systems."

asked; it also keeps a record of each student's score throughout the program. Probably several hundred[46] student responses could be easily returned to the transmission point, and the program could be stopped and discussed if large numbers of incorrect answers were coming in. In this way, the high quality that can be provided in programmed, recorded media can be supplemented by responsive on-the-spot discussion of troublesome points. Each student can also be apprised of his performance, and his questions can be dealt with by other students or the classroom teacher.

The cost of a digital terminal is relatively small, perhaps on the order of $30 at most. This is less than the cost of an audio microphone such as those used in the university level systems, and unlike the microphone, the digital terminal serves an individual student rather than an entire classroom. However, when the costs of the control console, projection control box, and other equipment are included, the initial cost of equipment for systems such as EDEX can run from $100 to $300 per student. The cost will, of course, depend on the configuration selected and must be judged in light of the services provided—immediate knowledge of the progress of the class as a whole and of each individual student, a permanent diagnostic record for each student, and administering and scoring tests.

As noted, such systems have been in use for several years; some have used videotape as a software medium. However, although there have been proposals to implement these systems using cable TV (in Philadelphia, for example), no digital response system with display and recording of student response has actually been tied into a cable TV system. Therefore, before such a system could be implemented, some engineering development would be needed.

Self-Directed Study. Self-instruction requires that the student have access to study materials at his own discretion, to meet needs as they arise and to follow individual interests. It does *not* necessitate on-line, adaptive programs such as those provided by computer-assisted instruction, but individual, random access to libraries of films, tapes, books, or other recorded materials is mandatory. Access may be manual or automated, depending on the number of programs available and the number of individuals requesting materials at any one time. The use of cable TV for self-instruction is currently economically feasible only when the cost of conventional instruction is high, for example, industrial and military training, continuing professional education, and graduate education.

Fully automated, completely independent, random access, in which any student can access any program in the library at any time, is expensive and at the cutting edge of the state of the art. Such a system must either contain as many copies of each program as will ever be used at one time or must be equipped to copy programs as they are called for so that a student can immediately access any program from its start, regardless of whether another student is already using it.

But less ambitious versions are useful. The conventional library or media center is the obvious example. Accessibility to programs (audio or audiovisual) stored at such repositories can be enhanced by providing at remote stations display equipment connected to the repository via telephone or television channels. A student at the remote terminal can request material either by voice or by a dial system, such as the one in operation at Oral Roberts University.[47] These systems do not provide

[46] The present console accepts up to 60 student responses. The largest system installed to date has a capacity of 660 students. These numbers and the following cost data were provided by the manufacturer.

[47] *Educational Product Report, No. 36, Dial Access Systems and Alternatives.*

independent access, however, because once a student has started using a given program, other students wishing to use the same program will either begin receiving it partway through or will have to wait until the program is not in use.

Another type of semiadaptive self-directed program, developed by Wilson,[48] consists of a taped lecture (but could be any style of presentation) supplemented by answers to a set of questions that may occur to the student as he is listening to or watching the presentation. After the presentation, the student consults a question map, a display of questions related to segments of the lecture. He then punches in codes for those questions that most closely match his own concerns and receives prerecorded discussions on those points. The technique requires careful preparation of both the lecture and the question map, with considerable time for pretrial with representative samples of students. Nevertheless, students report that it is a highly satisfactory study device and that they feel it provides more effective interaction with the teacher and the subject matter.

The lectures and answers to questions are recorded on audio tape, with drawings and other visual materials recorded on the second track of the tape by means of an electrowriter, making it possible for the professor to illustrate what he is saying in much the way he would use a blackboard. The electrowriter has two units: a transmitter, on which the speaker writes with a pen on a roll of paper, and a receiver in which a stylus reproduces the writing in ink on a similar roll of paper. It costs from $1500 to $4000, depending on the model chosen. The complete recording is packaged in audio cassettes, with answers to specific questions indicated by index codes. Using these codes, students retrieve answers to their questions manually. All equipment is available off-the-shelf. The major investment is clearly in the development and test of the recorded materials. Engineering development would also be required to adapt the system to cable TV.

Students who are deficient in study skills are likely to find programmed instruction more helpful for self-study. Linear programmed instruction presented in programmed text or on teaching machines helps the learner pace himself through the material, as he is aware of his progress at each step. A few systems are being developed at present that will provide on-line responsive programs over television channels. One of these, the TICCIT system, will use still-picture TV to increase channel capacity and thereby lower cost. The other, Plato IV, will use a plasma panel to store information at the terminal. Both require computers to supply the adaptive logic and are therefore considerably more expensive to install and program than the systems we have been discussing. Both systems are several years in the future.[49]

None of these systems, no matter how advanced, fully satisfies requirements for self-directed study, except for the academically sophisticated student. Even he, however, will prefer to have "internal random access" to study material, that is, to be able to skim some parts of the program, linger over others, and skip back and forth at will. At present, only printed materials give him this liberty; developments in film and tape cassettes will permit audiovisual programs to be used in this way in the future.

[48] "Interactive Lectures."
[49] Hammond, "Computer-Assisted Instruction."

SUMMARY

Cable TV is one of several ways to transmit audiovisual information. Local conditions will determine whether it is a feasible transmission medium and whether it is more practical than other available media. If the target audience is dispersed over an area whose radius is greater than 5 miles, interconnected cable systems or other media of transmission must be used. Also, if the target audience is at home and in the lower economic strata, special provisions will be needed to give them access to the cable system.

It should be emphasized that whatever media are chosen, they should be used only as means to solve pressing needs in education, not as ends in themselves. Applications of television that most obviously satisfy this requirement are those that reach audiences outside of regular school buildings—out-of-school uses. Probably the most significant educational uses of cable television will be these out-of-school uses.

Television can help meet several general needs in education; in fact, it can provide some services that can be provided in no other way. Cable television can be even more useful than other television media have been in the past because it offers additional technical features at a cost that is often below that of all but broadcast television.

Because schools have not made significant use of television in the past, there has been little of the development needed to apply its full potential to educational services. Most of this needed development is in the production of high-quality programming, although some, such as adaptive programming and indexing of recorded programming, requires new hardware as well.

The success of Sesame Street can be attributed in part to its use of the magazine format of commercial entertainment television. Thus, many of the techniques and talents for making television a useful teaching tool may already exist. What is lacking is the drive to apply them. Why this drive is lacking and what may be done about it are discussed in the next chapter.

III. BARRIERS TO THE USE OF TELEVISION IN EDUCATION

In this chapter we shall discuss the two major barriers to the use of television in education, the lack of high-quality television programming (software) suitable for school use and the inadequacy of means for using what programming does exist. We shall identify the likely sources of these obstacles and suggest some ways for surmounting them.

THE SOFTWARE PROBLEM

From several case studies of the use of TV in public schools in 1969, Wagner et al. concluded that the greatest single barrier to the use of TV is the lack of high-quality films, videotapes, and live programming. Most of the programs being used were of unknown teaching effectiveness because of

lack of formal, continuing, student-performance classroom evaluation procedures. Whatever their [teaching] quality, students and teachers subjectively felt that most existing programs were of poor quality. Undoubtedly, many were using commercial TV as a standard, but this single factor cannot account for the wide-spread impression of generally low ITV program quality in most school systems.

Poor program quality was found to be related to several factors. Basically, the "cottage industry" approach to ITV program production, characteristic of many of the school systems visited, in which it is envisioned that most of the ITV programs to be used by the system would be produced within the system, has been a failure to date.

Costs for local program production have been the single largest ITV operating budget line item. Even so, the resources have been inadequate for producing high quality programs because the resources have been spread too thinly over too many programs.

The relief from high local production costs that was anticipated through "program-sharing" among school systems has not occurred; program-sharing has run afoul of existing copyright laws. These contain a tangle of restrictions related to use permission and fees. Two key issues are: the residual rights of teachers involved in producing a program; the use permission and fee paid by a school system producing a program, when a proprietary clip is used in the program, not being applicable to the program's use by a "borrowing" school system.[1]

Some have argued that the programming should be produced by regular faculty because this will ensure teacher commitment to the project. The merits of this seemingly reasonable argument evaporate in real situations, however.

A dramatic case in point is the recent attempt by Scarborough College of the University of Toronto to fend off a growing shortage of teaching staff by making videotaped lectures the core of its curriculum. (The lack of teachers was predicted to reach such a level by 1972 that 10,000 university students would have to be turned away.) One of the key assumptions in the design of the TV college was that it would teach only the General Programs in Arts and Sciences, courses designed to produce a liberal arts education in three years. In the first year of the Programs, the choice of courses was very limited, and even in the upper years, options were few. Thus, by the time Scarborough reached its anticipated full enrollment of 5000 students, planners expected that as many as 1200 students would take first-year English; 1000, first-year mathematics; and 500, first-year history. Enrollments in other first-year courses were expected to be similar. According to Lee,

> These massive concentrations of students provided an ideal application for the technology of television teaching. Lectures could be prepared in advance and recorded on videotape, to be replayed during the academic year in lecture theatres holding about 200 students at a time. This would eliminate the need for enough faculty members to deliver the same content to several classes in a course, or the need for one professor to lecture to a class of 500 to 1,000 students. Since the content of these large introductory courses did not change substantially from year to year, the same tapes with slight revisions could be replayed for several years.
>
> This arrangement was expected to allow for 30 per cent fewer teaching staff than a comparable "live" system would need to handle the same number of students. Not only would teaching salaries be saved, but also the office space for these teachers. The secretaries, cleaners, and other support staff required by teachers would also not be needed. These savings would pay both the capital and operating costs of television facilities and would produce a surplus over the comparable cost of "live" instruction of about $1 million a year when the 5,000 enrolment was reached.[2]

The college very nearly foundered within a few months of its inception because of faculty disenchantment with TV. Although resentment and anxiety appeared to center on legal questions concerning the production of videotaped lectures, a transcript of a special meeting on the problem shows that many of the faculty were in effect defending their traditional method of teaching against a threatening innovation:

> At one point, this debate became direct and explicit. One professor, referring to the process of developing a television lecture program with script, visuals, studio rehearsal, and editing asked: "How essential is this whole production treatment?" The professor explained that his method of lecturing was simply to walk into the classroom and start talking. . . .[3]

[1] *A Study of Systemic Resistances to Utilization of ITV in Public School Systems.*
[2] *Test Pattern.*
[3] Ibid.

Scarborough planners had not thought far enough ahead to do research on the effects of their proposed system on the faculty and students. Rather, they assumed that TV would, somehow, carry the day. They ignored two essential facts: (1) professors, by nature, are not inclined, and often not able, to change their customary methods of teaching, and (2) the students, who are the TV generation, expect much more of television than a college can afford.

Traditionally, universities have shown much less concern for a faculty member's teaching ability than for his ability to publish. This had special impact on Scarborough College teachers, as Lee points out:

> There is one important factor which makes Scarborough College teachers less likely . . . to be willing to take the risks involved in developing television skills. They are frequently concerned with being out of the mainstream of activities in their departments, set apart in something of a remote outpost. Publishing becomes even more than normally important, to ensure the attention of the colleagues downtown *who decide on promotions*. Making good television is rarely counted as publishing.[4]

Finally, the TV lecturers could see that they had more at stake in producing a videotape than in giving a live lecture. They were on record for anyone to see and quote. In the studio their effectiveness was on display before the technical staff. And on videotape, their ability to use television—where the camera eye can magnify ordinary teaching deficiencies—was readily apparent to the students.

As for the students, they demanded the invitation to involvement that is built into much commercial television, but which many educators reject as "mere entertainment." And, like most university teachers, Scarborough planners assumed that lectures were more important to the students' learning than they actually were.

The Scarborough experience is not unique. From their case studies, Wagner et al. concluded that

> Attempts to insure high classroom utilization of locally produced programs through "involvement" of classroom teachers in the production process have been counter-productive in terms of both program quality and teacher acceptance. Most classroom teachers have neither the desire nor talent to be so involved; there seems to be little positive transfer from the ability of some teachers to recognize a "good" program to an ability and willingness to produce one. Agreement on program specifications is difficult to achieve among teachers themselves; teacher views frequently clash with professional production views. The inevitable compromises in content and pedagogical technique necessary in such situations frequently results in a program judged by even the teachers who were involved to be of poor quality. The combination of frustration in "involvement" and dissatisfaction with end-product clearly militates against high utilization, particularly when an objective evaluation system, based on student performance, is not available.[5]

Initially, there was little or no programming of usable quality available to the schools. Even instructional films were either of poor quality or had been made so general in their appeal (to prorate the cost of quality production over a large audience) that they did not fit into specific school curricula. Production of high-quality

---

[4] Emphasis added. Ibid.
[5] *A Study of Systemic Resistances.*

programming tailored to localized needs can be prohibitively expensive for small school systems. Even simple programming, which is unlikely to interest today's students or teachers, can cost from $25 to $100 an hour, and more elaborate programming will cost considerably more.[6] Sesame Street, for example, cost $30,000 an hour during its first year of production.

Of course, Sesame Street costs were compensated by the program's large-scale use. During the last few years, the National Instructional Television Center (NIT)[7] has promoted and coordinated the establishment of consortia for producing education series for large-scale use on ITV. For example, the popular Ripples was planned and financed by a group of experts on early childhood and of television specialists from fourteen education agencies. The Northern Virginia Educational Television Association produced the programs. Many of the participating agencies were state-based, although several educational television stations associated with larger school systems (such as WETA in Washington, D.C.) or universities were included. Ripples cost $12,000 an hour and was two years in planning and production. A 31-member consortium is currently producing a series in health education (Inside Out) that will cost $90,000 an hour.[8]

Such ventures have resulted in high-quality programming and a decrease in the percentage of locally produced programming transmitted by broadcast ITV. This is illustrated on Fig. 3, taken from Rockman.[9] Another result has been support of the consortium concept by the Council of Chief State School Officers and promotion of a permanent national organization for developing quality school television programming.[10]

None of these series is intended for basic instruction, that is, to teach subjects in a regular school curriculum. They are produced for broadcast TV and are to supplement or enrich basic instruction. For example, Ripples is intended to help children build human values, extend their knowledge, increase their aesthetic sensitivity, and understand the changing nature of the real world.[11] However, when most teachers are supplied with high-quality programming in the basic curriculum such as Sesame Street and the Electric Company, they are happy to use it in their classrooms. But this attitude would probably not persist if the costs of producing such programming were defrayed by reductions in the cost of other parts of school operations rather than simply added to other costs, as at present.

How can a local school system support the costs of programming tailored to local needs? The best advice is to share resources for program origination with another group. Baer[12] lists the following possibilities:

- Use the cable operator's studio and equipment.
- Form a new consortium of education groups.
- Use existing school facilities.

[6] Baer, op. cit.
[7] Other sources of ITV programming are listed in Part IV.
[8] National Instructional Television Center, *Newsletter,* September/October 1972.
[9] *One Week of Educational Television.*
[10] *Education Daily,* November 21, 1972.
[11] National Instructional Television Center, *Ripples.*
[12] Baer, op. cit.

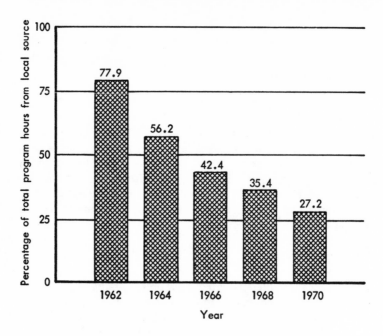

Fig. 3—Broadcast ITV local production, 1962-1970

A number of cable operators have arrangements with schools for the use of facilities and equipment. Also, local consortia have been or are being established for sharing the costs of educational programming. (Both of these options are discussed further in the next chapter.) Finally, a number of schools, such as Pennsylvania State University (WPSX-TV) and the University of Southern California, already have studios for producing their own programming. Schools with their own facilities might be willing to help other education agencies with program production.

Not only is high-quality programming lacking, but there is almost no way to find out whether programming exists that will meet a specific educational need. The repositories of instructional programming do not provide search or rating services; their catalogues usually give only the general subject area and grade level of the prospective student. There are no lists of instructional objectives pursued by the programs, specific delineation of the target audience, evidence on teaching effectiveness, or indicators of quality.

An objective rating board, unaffiliated with a production agency or other group with vested interest, should be established to search for, test, and rate instructional television programming, in somewhat the way that nonprofit agencies rate consumer products. The establishment of such a board is an appropriate task for the new National Institute of Education.

THE UTILIZATION PROBLEM

Utilization activities fall into four major categories:

- Concurrent instructional activities that integrate the televised programming into the school curriculum.
- Audience monitoring (of both students and teachers) to determine the effectiveness of the programming.
- Development and maintenance of adequate facilities for distributing TV programming.
- Publicizing the program to the intended audience (for out-of-school programs).

In their conclusions from the case studies, Wagner et al. also noted that

ITV classroom utilization, as a process independent of production and transmission, has received an inadequate allocation of resources in school systems. More often than not

- classroom utilization aids,
- adequately compensated teacher training,
- equipment maintenance,
- objective classroom program evaluation, and
- formal utilization feedback systems

were, if existent, insufficiently programmed and supported in the school systems to have had a significant impact on ITV utilization rates.[13]

Therefore, they recommended that school systems shift their emphasis from program production to program utilization.

Although utilization activities receive short shrift in most schools, there are examples of good utilization efforts. In Anaheim, for example, instructional committees cochaired by television teachers and classroom teachers determine *what* is televised, *when* it is televised, *how* it is televised, and what the evaluation shall consist of. These committees operate at each grade level for each subject area taught. The process includes setting objectives as well. Classroom teachers also participate, along with television teachers and producer-directors, in developing lesson guides.[14] During its first year of operation, the Children's Television Workshop (CTW) spent $700,000[15] on utilization, working through 35 field coordinators assigned to 12 metropolitan areas.[16] These people set up viewing centers in Head Start facilities and other public areas. CTW also spent $400,000 on promotion through the mass media and by means of special materials.

Careful attention in two areas will help ensure that the televised programming is utilized effectively:

---

[13] *A Study of Systemic Resistances.* The Appendix indicates that, in addition, few schools had near enough TV sets for routine use of TV. Only 1 out of 10 schools had as many as 10 sets in classrooms. (With an average of about 22 classrooms per school, 10 classrooms with sets would be comparable with the TV classroom setup provided in TV-intensive systems such as Hagerstown and Anaheim. These systems, however, generally have more than one set in a TV classroom.)

[14] Anaheim City School District, "Teaching with Television."

[15] Connell, "How Sesame Street Raised and Spent Eight Million Dollars."

[16] *Scope,* Field Services Department, Children's Television Workshop, New York, June 1971.

1. Utilization activities should be planned systematically from the outset. These include designing the TV distribution system and its monitoring and maintenance, structuring concurrent instructional activities, planning the production and distribution of supporting instructional materials, planning teacher training, and designing the program monitoring and evaluation components.[17]

2. The needs, interests, and wishes of the prospective audience should be determined as definitively as possible and consulted from the beginning and throughout the life of the program. (For in-school programs, the audience includes the classroom teachers.) Representatives of the audience should participate in program planning and development. Teachers should be trained to use the televised programming in realistic workshops and to conduct concurrent instructional activities. In-service training should be provided routinely after the program has been established.

THE BASIC PROBLEM

Lack of appropriate, high-quality programming and inadequate provisions for its utilization are not the primary barriers to the use of television in education. Rather, they are symptoms of a pervasive problem—lack of interest in or outright resistance to major change by the education community. The propensity of education institutions to maintain the status quo has been observed and studied for a number of years. The primary cause is behavioral in nature, as suggested by the Scarborough College experience. Educators, whether faculty or administration, find it difficult to change their concept of what the process of education should be. Even when the teacher in the classroom is actually replaced by television, it is usually a surrogate classroom teacher who replaces him. Very few programs have broken far enough out of this restrictive mold to achieve even part of the potential for instruction offered by the television medium. And even when the programming is stimulating, if it is only an adjunct to the teacher's business as usual, the production expense is hard to justify.

If television is to be more than an adjunct, the teacher must take on the new roles of coordinator and manager and submerge the old role of primary purveyor of instruction. Since teaching is a person-to-person activity, it is a direct expression of the teacher's personality, and like other human beings teachers are reluctant to change their image. The profound revision in teaching style that would be required of a teacher making full use of television can be brought about and maintained only by a continuous, concerted effort. Teachers must be given incentives to change their ways and must be taught how to do so. The incentives will be harder to come by than the training. Monetary incentives buck the teacher unions with their opposition to merit pay. Teachers who do an outstanding job of utilization might be given rewards tailored to their own values. Special activities for their classes or professional recognition might also be effective.

Other factors exacerbate the situation. The two that are the most difficult to deal with are closely interrelated. One is the complexity of the process of education;

---

[17] These activities are described in detail in Part IV.

the other is the lack of usable information on effective change. Teachers, administrators, students, and the community interact in complex and subtle ways. Materials, methods, facilities, equipment must be appropriately fitted together to support these human interactions. This complexity can turn the development and implementation of changes in the education process into a nightmare. It is little wonder that most school administrators prefer to live with processes that at least run smoothly and that run themselves to a large extent. The companion volume is aimed at easing problems of development and implementation by setting forth steps in planning and implementation in detail.

Complexity also makes it hard to define change in terms that communicate readily. Changes are often described by means of catch phrases (such as *individualized instruction)* that tell the educator nothing about how to implement the change in his schools. Evaluations are often thinly disguised sales pitches that leave the educator distrustful of all miracle solutions to educational problems. Even honest evaluations rarely treat more than one or two outcomes, whereas education projects affect their participants in many ways. Evaluations almost never describe the programs in detail, let alone recount the steps needed for successful implementation. The companion volume also treats these problems in the chapter on evaluation.

IV. THE EDUCATOR'S STAKE IN THE CABLE TV FRANCHISE

How well a cable TV system will meet the needs of the education community depends not only on the community's initiative but also on the arrangements that are made for cable service to the schools. These arrangements are usually (but not always) spelled out in the franchise under which the cable TV system will be operated. Provisions of greatest concern are those regulating

- The administration of the free education channel mandated by the FCC.
- Access to additional channels for lease and reasonable leasing rates.
- Connection of school buildings to regular cable service and reasonable installation and monthly charges for that service.
- Charges for initial installation of multiple connections within school buildings and additional service charges for buildings with multiple connections.
- Providing or sharing facilities and equipment for production and transmission of educational programming.
- An early date for operational two-way capacity.
- Incorporation of new technical standards when they are set by governmental agencies.

The discussion of these issues will draw, where relevant, from current practice, the recent rulings of the FCC, and policy statements by several influential groups concerned with the use of television in education.

We have said that the FCC has required that newly franchised operators in the 100 largest television markets provide one free channel for educational use during an experimental period of at least five years. Systems franchised before March 31, 1972, will have at most five years to comply with this requirement. Other possibilities, which may or may not be stated in the franchise, should also be considered. For example, it may be possible to negotiate the lease of additional channels or to use facilities that are not needed full-time by the cable operator. The operator may be eager to cooperate to increase his output of locally originated programming, to attract subscribers. And more than a single channel may be made available free to the education community if a special showing of need and plans for use can be made to the FCC.[1]

[1] See Baer, op. cit.

The implications of the FCC ruling for preferential treatment of community groups is unclear. Right now, the FCC seems to view 3 to 5 percent of the cable system revenues as a nominal maximum return to the franchisor. Any franchise provisions that would, in effect, grant the franchisor a higher return will probably be treated as potentially burdensome to the franchisee and will require a special showing.

The FCC is much more likely to consider requests for preferential provisos favorably if the showing supports the need with a viable plan and demonstrates the ability of the cable system to provide what is asked for. Thus, the importance of careful planning by those who view cable TV as a way to improve education cannot be overemphasized. Only after such people have argued to the FCC a series of well-supported cases will the rulings be clarified and will it be possible for the education community to take for granted certain forms of preference. All of the possibilities for preferential treatment discussed in this chapter should be viewed in that light.

Of course, preferential treatment that adds no financial burden to the cable operator can be provided without requiring a special showing before the FCC.

FRANCHISE NEGOTIATION

The FCC requires that local franchise proceedings include "public hearing affording due process." Representatives of the local school district, institutions of higher education, and other institutions concerned with preschool or adult education should participate in these hearings. After the franchise is awarded, the same group should promote the establishment of a local regulatory agency that will protect the public's interest in the system.

As a general rule, the franchisor is a body of a municipal government, such as a city council; the franchisee is a private, profitmaking concern. But these are not the only options, and in some instances, arrangements such as municipal or nonprofit ownership have been considered. The prime argument for such arrangements has been that the needs of the community or of special community groups must not be subordinated to the cable TV system operator's desire for profits. In its policy statement to the FCC on cable television, the National Education Association (NEA) strongly supported such arrangements:

> in an effort to guarantee the public a fair share in public cable communications, the Commission should encourage experimentation in selected local communities with the development of public cable corporations dedicated to fostering a richly beneficial system serving the public interest.
>
> Schools might well take the leadership in local communities in mobilizing public institutions to form such public cable corporations which themselves could operate the local franchise in a given community. Such cable corporations should include not only the public broadcasting stations but also public schools and other public education organizations and institutions. In some communities a public nonprofit cable authority (similar to a port or power authority) may be a more appropriate licensee of the noncom-

mercial CATV franchise in that community than the public broadcasting station.[2]

Since the FCC has now required cable TV systems to provide three free channels, one for educational, one for local governmental, and one for public access use,[3] there may be less interest in public ownership in the future. Most communities, however, should still consider nonprofit or government ownership of the cable TV system before awarding their franchises.[4]

ADMINISTERING THE EDUCATION CHANNEL

The NEA recommends that beyond the three access channels additional channels "be provided free of charge up to a total of 20 percent of the system's capacity for educational, instructional, governmental, and public access programming, *as the schools and the public demonstrate their ability to use these channels.*"[5] The NEA has based its recommendation on precedent, noting that

> In the 1940's, public policy dictated that 20 percent of FM radio frequencies be set aside for educational and other noncommercial uses. In the early 1950's, when new broadcast television channels were established, approximately 20 percent were likewise reserved for educational use.[6]

But most people who have FM radios and UHF television receivers are well aware that public agencies have not begun to make use of the frequencies allocated to them under these rules. Thus, the precedent has to do with rules, not with usage.

The recent FCC ruling is more in line with actual usage than is the NEA position. At present, almost all cable TV systems providing channels to the schools supply only one and it is free, according to a recent survey of cable TV operators by the National Cable Television Association. The ruling does not, however, meet the needs of school districts contemplating the use of TV for distribution of films and other audiovisual materials on demand. Because of the requirements for scheduling, this use alone can easily consume several channels, as demonstrated by the Ottawa project. It also does not satisfy the needs of schools wishing to use television for basic instruction above the elementary level. Again, difficulties of scheduling can impose a requirement for over ten channels.[7]

With regard to the education channel, the regulations require only that it be made available by the operator free of charge for an experimental period of 5 years, that advertising, lotteries, and obscene and indecent material be barred, and that a list of applicants for the channel be maintained for two years. Moreover, the rules debar local agencies from establishing additional rules governing the education access channel:

[2] Wigren, "The NEA's Position on Cable Television."

[3] FCC, "Cable Television Report and Order."

[4] See NEA, *Cable Television: Franchise Provisions for Schools,* and Baer, *Cable Television: A Handbook for Decisionmaking* (Chap. 4), for a discussion of various ownership alternatives.

[5] NEA, *Cable Television.*

[6] Wigren, "The NEA's Position on Cable Television."

[7] Bretz, "The Potential Uses of Cable."

> Except on specific authorization, or with respect to the operation of the local government access channel, no local authority shall prescribe any other rules concerning the number or manner of operation of access channels. . . .[8]

There is a real danger that the cable operator may restrict the channel to a single school agency to simplify management or for some other reason. In some cases, it would be desirable to establish a separate entity comprising several users with jurisdiction over the education channel to assure that the needs of all potential users are considered fairly and to spread the cost of expensive items such as local program production, computers, and software repositories. Establishment of such an entity would require a special showing to the FCC if the group were to have any say over the operation of the channel, and established education institutions may be reluctant to give up their traditional autonomy in the areas of curriculum and teaching practice—precisely the areas in which television would have a major impact.

Even so, joint arrangements have been considered in a few cases. Worth noting (although not associated with a cable system) is the Educational Television Association of Metropolitan Cleveland (ETAMC), with members drawn from the Cleveland education community. Some of the participants are the Diocese of Cleveland, the Board of Education of the Cleveland City Schools, and the Board of Education of the Parma City School District. ETAMC is an umbrella agency that has been awarded 16 ITFS channels to serve its members. In making the grant, the FCC waived the section of its rules which provides that a licensee is limited to four channels in a single area of operation. The community television station, WVIZ, is also a member and provides the other members with nondiscriminatory services at published rates and under the approval of the FCC. WVIZ has been in the forefront of instructional television and has produced many instructional series that are now distributed nationally.

ETAMC is controlled by a board elected by the members. The board is responsible for planning the allocation and use of channels, scheduling time, coordination and use of programming, establishment of fees and charges, and all other administrative functions:

> WVIZ will operate and maintain the ITFS transmission equipment and will provide studio space, if needed, to individual users. The individual school systems and organizations participating in the ITFS operation will subscribe to it, but each may also provide its own programming or share in materials transmitted by other systems. Costs of local programming and distribution and charges for printed resource materials will be borne by the developing agency. No one group will have a channel, but each will be allocated time, determined by the governing board in accordance with the party's request and with the amount of time available. The only requirement placed by ETAMC on member groups is that the programming be of minimum broadcast quality. The ITFS system will be operated as complement to broadcast Channel 25. Since the two systems will not be competitive, participating schools will be able to use both ETV and ITFS programming.[9]

---

[8] FCC, "Cable Television Report and Order."
[9] *Educational Product Report, No. 31,* p. 31.

At present, ETAMC serves 53 systems representing nearly half a million students with five operating channels.

A similar arrangement has been set up to coordinate the development of cable TV systems in Orange County, California. The Public Cable Television Authority, a joint powers authority of five cities, has been established to plan the development of a regional approach to cable TV. The board of directors comprises an elected official from each city. The PCTA is now at the point where specifications for the system can be released to franchise applicants.

Various joint agreements are currently being drawn up among other school districts. Three cities in California (Ontario, Montclair, and Upland) are sharing a cable TV system through separate franchises with the cable company. All such arrangements will require special showings, as noted previously.

The NEA has recommended that "a cable educational advisory board be established in every community as part of the franchise." This board would set policy for the use of the education channel in a number of respects and would develop some educational programming.[10] Baer has urged, in discussing a similar recommendation, that

> This Board should not be dominated by public school authorities or any other single educational body. It might include elected school board officials, public school teachers and administrators, private school, community college, and university representatives, and private citizens unaffiliated with any educational group. The cable operator could then administer the rules for access set by the Board.[11]

In Chap. 2, we suggested that the most significant use of cable TV in education will require establishing new organizations and institutions; within-school uses will be of secondary importance. This underscores the need for decisions on the use of the education channel by a coalition of members of the education community.

The FCC ruling avoids restricting the users of the education channel to *established* education agencies, although users must be "local educational authorities."[12] This may leave the door open for the establishment of new institutions to use the cable TV system.

Ideally, the education channel will be allocated to those who can make best use of it in terms of need and cost. Criteria for selection of users should include

- The number in each potential audience.
- The importance of the potential use of the cable TV system.
- The extent to which the cable TV system provides services that cannot be provided better—if at all—in some other way.
- The cost *(including the cost to the audience)* of providing the educational service in some other way.

Guidelines for performing such analyses are discussed in the companion volume, R-1144-NSF.

To answer these questions, potential users should be partitioned between daytime and evening hours. During daytime hours, the largest potential audiences are

[10] NEA, *Cable Television.*
[11] Baer, op. cit.
[12] "Cable Television Report and Order."

in schools or other service institutions or at home. For evening hours, programs for working adults or college students are most appropriate.

USE OF THE OTHER FREE CHANNELS FOR EDUCATION

A case can be made for using the channels set aside for local governmental agencies and for public access for educational applications, but the case is not open-and-shut. Training for municipal employees, public forums concerning school problems, programs dealing with consumer dissatisfaction, public expressions of dissatisfactions of minority groups with the schools or the local government are all subjects for programming falling into gray areas between the free channels. The determination of what agencies can use the free channels for what purposes is one of the pressing policy issues to be faced by both potential users and the FCC. It will be resolved only by direct test in specific situations.

LEASED CHANNELS

According to the FCC rules, the franchising authority must obtain FCC approval before requiring the provision of channels to the schools free or on a preferred basis:

> Because of the federal concern, local entities will not be permitted, absent a special showing, to require the channels be assigned for purposes other than those specified above. We stress again that we are entering into an experimental or developmental period. Thus, where the cable operator and franchising authority wish to experiment by providing additional channel capacity for such purposes as public, educational, and government access —on a free basis or at *reduced charges*—we will entertain petitions and consider the appropriateness of authorizing such experiments, to gain further insight and to guide further courses of action.[13]

On the other hand, the FCC rules *require* cable operators to lease additional channels as needed. Therefore, education agencies that need more than the free channel can lease them, perhaps at a preferential rate. For example, a charge could be made

> covering at least the incremental or out-of-pocket cost of supplying the channels, but . . . *less* than a proportionate share of the overhead of the cable operation in comparison with the charges for other services. . . . Such a pricing practice, in which some services bear much more of the overhead than others, is common in virtually all industries.[14]

As long as *all* users pay more than the cable operator's marginal cost of providing a leased channel, preferential rates are not a burden. Baer suggests

[13] Emphasis added. Ibid.
[14] Baer, op. cit.

One approach would be to set multi-part rates, one part of which would be proportional to the revenue that the lessee receives from channel use. An educational user who gains no revenue from the leased channel might pay a base price of, say, $30 an hour, while a pay-TV promoter might pay more than $100 an hour.[15]

We have already noted that many of the uses for television in education can easily require considerably more than one channel. This is especially true if an attempt is made to tailor transmissions to the needs and interests of special audiences. In addition, a number of education agencies now use ITFS for transmission of educational programming. The NEA recommends that "The franchisee be required to connect with the master control of any ITFS system in the franchise area, as requested. . . ."[16]

SCOPE,[17] a nonprofit education agency serving 72 school districts on Long Island, would go further in its model franchise:

> the company must provide . . . distribution of one ITFS channel during the first five years of its franchise. The second five years, the company must provide a second channel, and during the third five years, a third channel.

Such a provision would probably require a special showing to the FCC. The commitment to educational uses of television demonstrated by the community's willingness to assume the effort and expense of installing its own ITFS system would provide strong support to the showing.

At the least, education agencies should have a guarantee that as cable leasing time becomes available they will have access to it. For example, ERDC has urged that educators have the right of first refusal on additional channels;[18] this may not require FCC approval. Whatever can be worked out, the franchise should make explicit provisions for access to increased channel capacity by the education community and for rates for leasing time.

OTHER PREFERENTIAL ARRANGEMENTS

According to the survey conducted by the National Cable Television Association in 1972, cable operators routinely give preferential treatment to the education community in a number of ways. For example, of the 85 percent of the operators who provide regular service to schools, 70 percent[19] provide free connection of schools to the cable system. Other aspects of this service are shown in Table 3.

With regard to connecting schools to the cable system, the NEA recommends that schools within 100 yards of the trunk line should be connected to it free and schools beyond 100 yards at cost, and that no monthly charge should be made for

---

[15] Baer, op. cit.

[16] NEA, *Cable Television.*

[17] SCOPE: Suffolk County Organization for the Promotion of Education. The quote that follows is from an article by SCOPE's director Roger W. Hill, Jr., "Educational Considerations of CATV Cablecasting and Telecommunications."

[18] Shafer, *A Cable TV Guide.*

[19] Figures have been rounded to the nearest 5 percent. The preliminary response rate to the questionnaire was about 30 percent, so finer accuracy is probably not justified.

Table 3

CHARGES OF CABLE OPERATORS WHO PROVIDE REGULAR
SERVICE TO THE SCHOOLS

Service	Charge for Service (percent of operators)		
	Free	Less than Standard	Standard
Installation	70	25	5
Monthly service	90	5	5
Intrabuilding installation	30	40	30
Monthly service for multiple connections	95	5	0

regular cable service to the schools. With regard to intrabuilding connections, the NEA recommends that the cable operator should provide these at cost, since "to request more would be an unreasonable financial demand upon some cable operators."[20] These requirements are not far out of line with current practice. The FCC is silent on these points.

Only 30 percent of the cable operators reported that they charged schools the full rate for making multiple connections in buildings. We do not know, however, how many connections were made, on the average. It is surely less than all rooms, as even in school systems making heavy use of television (such as Hagerstown) only half of the rooms are connected.

Twenty-five percent of the cable TV operators reported that they made studio facilities or equipment available to the schools for production of school programming. Of these, 75 percent allowed schools to use the facilities or equipment without charge. If a large school population will use the cable system, the franchise should contain provisions for production arrangements suitable to the school's needs. NEA suggests that such arrangements are best made through mutual agreement at the local level. Although the FCC rules require the cable operator to provide production facilities for public access, he is not *required* to extend the same service to users of the education channel. "Still, cable operators want to encourage new applications, and they will generally be receptive to aiding school . . . users. The amount charged for the use of facilities would be worked out in each individual case."[21] NEA believes that the education community will be able to use the cable operator's studio facilities under the same arrangements as users of the public access channel.

[20] NEA, *Cable Television.*
[21] Baer, op. cit.

TWO-WAY USES

The FCC now requires that a "capacity" for nonvoice return communications be built into new cable systems in the major television markets. This means only that the system should eventually be able to provide return communications without "time-consuming and costly system rebuilding." Thus, most systems will probably not install even the capacity for return transmissions. If any services are provided, they will be low-cost data return links; these would be sufficient for such commercial applications as surveys, marketing services, and burglar alarm devices. Similar terminals could be used for instruction calling for selected response from the student, but educators should not expect to use this capacity for 5 to 10 years unless they will pay for the terminals themselves.

The NEA urges that when an education agency has devised a "viable plan for actual use" of two-way services more elaborate than simple data links, the cable operator be required to expand his system to provide these services:

> The NEA feels that this is a very important requirement for schools, inasmuch as cable's most unique feature is its ability to make possible interaction between teacher and learner and between schools in widely separated locations. If cable were not to provide this feature, much of its attraction to the educational community would be lost.[22]

With this statement, the NEA appears to base much of its support for cable TV on its potential for spatial expansion of the teacher-student interaction. Although this would be a worthwhile objective in many situations, the argument overlooks the other features of cable TV that make it attractive for education.

TECHNICAL QUALITY

The FCC has set minimum technical standards for broadcast television signals carried by cable TV systems, and the operator must certify, annually, that his system meets these standards. There are, however, a number of older cable systems that deliver signals that would be unacceptable for color reception today. Subscribers who can receive broadcast television signals only via cable may tolerate such service, but this seems poor justification. In addition, no standards at all are set for nonbroadcast services. The franchisor is permitted to specify more stringent standards if he will enforce them himself.[23]

At the least, the franchise should contain provisions that will require the incorporation of new standards as they are adopted:

> The rapid development of cable technology and new services makes it likely that a number of new standards will be set by federal and state authorities during the initial franchise period.[24]

[22] NEA, *Cable Television.*
[23] An example of desirable standards is given in Fink, "Public Education's Stake in the Proposed Community Antenna System Franchise," p. 51.
[24] Baer, op. cit.

SUMMARY

If the franchise for a cable TV system has not been granted, members of the education community can influence the negotiations in their favor. They can

1. Consider forming consortia with other users or using interconnected systems to share the costs of programming, computer terminals, and the like.
2. Ensure that the franchise agreement makes specific provisions, as listed on p. 32.

Provisions that, in effect, return more than 3 to 5 percent of system revenues to the franchisor will require a special showing to the FCC for approval. In fact, the FCC's new rules specifically enjoin the franchisor from requiring cable operators to provide a large number of free channels or other preferential treatment to the education community. On the other hand, it is likely that the FCC will approve exceptions to its rules if education agencies can make a convincing case that (1) they will make significant use of the cable system (that is, that the criteria stated by the NAEB are satisfied) and (2) whatever preferential treatment they request will not unduly burden the cable operator. The test of significance will usually be more difficult to fulfill than will the showing of financial feasibility.

ASSISTANCE IN FRANCHISE NEGOTIATIONS

Because many education agencies lack personnel who are familiar with the technical and legal aspects of cable TV franchising, a number of agencies have formed to assist schools in taking advantage of this new technology. Some of these agencies are listed below:

PubliCable, National Education Association, Harold Wigren, 1201 Sixteenth Street, N.W., Washington, D.C. 20036. Telephone: (202) 833-4120. Active in assisting educators in using cable.

Joint Council on Educational Telecommunications, Frank W. Norwood, 1126 Sixteenth Street, N.W., Washington, D.C. 20036. Telephone: (202) 659-9740. Has information on cable developments in education.

Suffolk County Organization for the Promotion of Education, Roger W. Hill, Jr., Suffolk Educational Center, Stony Brook, New York 11790. Telephone: (516) 751-8500. Will advise on franchise negotiations for educational applications.

Office of Communication, United Church of Christ, 289 Park Avenue South, New York, New York 10010. Active in encouraging citizen participation in the communication field, particularly of minority groups. Will provide free literature on request.

Appendix[1]

BASIC STATISTICS ON INSTRUCTIONAL TELEVISION AND OTHER TECHNOLOGIES

Three Out of Four Public Schools Now Have TV Receivers and One in Four Has Videotape Recorders (VTRs)

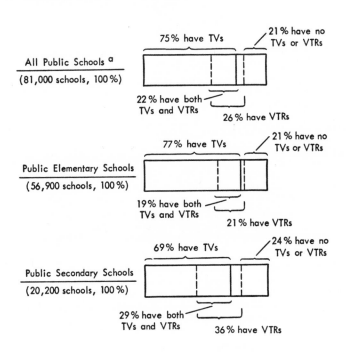

ᵃThe total for all schools includes an estimated 3900 combined elementary-secondary schools which are not included in data by school level.

---

[1]This appendix is reprinted from HEW, *Bulletin*, No. 7, February 9, 1971. The source of the material is the School Staffing Survey, Spring 1970. The survey was conducted by the National Center for Educational

Television receivers are more widespread among public schools in the United States for classroom use than either radios or videotape recorders according to principals' reports in the spring of 1970. Relatively few schools (9%) have closed-circuit/ITFS (Instructional Television Fixed Service) systems, but 11% of local public school pupils (which totaled about 44 million) are in those schools. A larger proportion of elementary than secondary schools have TV receivers, while the reverse is true for videotape recorders. See Table 1.

Table 1

DISTRIBUTION OF INSTRUCTIONAL TECHNOLOGIES

| Instructional Technologies | Number and Percent of Schools | | | | | |
| | All Schools[a] | | Elementary | | Secondary | |
	No.	%	No.	%	No.	%
Total schools	81,000	100	56,900	100	20,200	100
With TV receivers	61,000	75	43,700	77	14,200	70
With closed-circuit/ ITFS system[b]	7,100	9	4,300	8	2,800	14
With radios	44,700	55	27,200	52	12,600	62
With videotape recorders	20,900	26	12,100	21	7,200	36

[a]The total for all schools includes an estimated 3900 combined elementary-secondary schools which are not included in data by school level.

[b]Closed-circuit/ITFS television system refers to broadcasting not receivable by the general public. This is televised instruction or information only made available to specified locations for specific individuals or groups. The use of a combination of a portable videotape recorder and/or accompanying single camera and television monitor sometimes used for observation or magnification purposes is also considered closed-circuit TV.

Statistics, USOE, with technical assistance on instructional technology from the Bureau of Libraries and Educational Technology, and with support from the Bureau of Educational Personnel Development and the Bureau of Education for the Handicapped. The National Center for Educational Research and Development supported the pilot survey which preceded the 1970 survey. The data presented in this *Bulletin* are preliminary. This is the first time these statistics have been gathered. No trends can be projected from them.

The spring 1970 survey sample consisted of approximately 1200 public elementary and secondary schools, representative of the nation's public schools by three strata: large cities (over 100,000 population

Table 1 presents the following highlights:

o 82% of all pupils are in schools having TV receivers.

o In large cities, more than 90% of public schools have TV receivers.

o In suburban areas surrounding these large cities, 47% of secondary schools, with 56% of secondary pupils, have videotape recorders.

o Only 13% of schools, mostly away from large cities, have none of the above forms of advanced instructional technology.

More Than 70% of Schools Having TV Receivers Use Educational TV

Telecasts from educational (noncommercial) television stations are used by 53% of all schools, or more than 70% of the schools that have TV receivers (which is 75% of all schools). In terms of pupils, about 57% of the nation's public elementary and secondary school children are in schools which use educational telecasts. More elementary schools than secondary schools report that they use educational telecasts. Proportionately, more schools in large cities use educational telecasts than either the metropolitan areas surrounding these cities or other areas of the country. The highest proportion of schools reporting use of educational telecasts is 90% for large-city elementary schools. See Table 2.

Few Schools Report Many Units Permitting Simultaneous Instructional Use

Among all schools, the median number of TV receivers provided for classroom use[2] is 2, and the median number of radios is 1. Both of

---

as of 1960 Census), the metropolitan areas surrounding these cities, and all other areas of the 50 states and the District of Columbia. Although the data presented here are based on a sample, they are not likely to differ by more than a few percentage points from the results that would have been obtained from a complete canvass of all public elementary and secondary schools. Specific measures of sampling reliability of the estimates will appear in a forthcoming report of the survey.

For further information, contact Ronald J. Pedone of NCES's Statistical Development Staff who prepared the *Bulletin*; telephone: (202) 963-3170.

[2]Principals were asked to report sets available to any of the staff for classroom use, excluding any sets owned by staff members or pupils.

Table 2

USE OF EDUCATIONAL TELECASTS

| Category of Schools | Schools (no.) | | Percent with TV Using Educational Telecasts |
	Total	With TV Receivers	
All public schools[a]	81,000	61,000	71
All public elementary	56,900	43,700	75
All public secondary	20,200	14,200	59
Large city public	11,500	10,500	86
Metropolitan areas of large cities	22,600	19,200	72
All other areas	46,900	31,300	64

[a]The total for all schools includes an estimated 3900 combined elementary-secondary schools which are not included in data by school level.

these figures are influenced by the sizable proportions of schools reporting no TVs and reporting no radios.

Considering only the schools reporting at least one TV receiver, the computed median is 2.8; considering only those schools reporting at least one radio, the median is 1.6. The average number of pupils per school is 539 for all public schools--420 for all elementary schools and 852 for all secondary schools.

It is apparent that simultaneous use of television by different classes is necessarily very limited, at least for many schools. Further indications on this are given by the frequency distributions of schools and pupils by number of TV receivers available for classroom use, as shown in Table 3.

Three of four schools--accounting for 70% of all pupils--have four television receivers or less for classroom use. However, one of ten schools--accounting for 15% of all pupils--have 10 or more television receivers available for classroom use.

Table 3

NUMBER OF TELEVISION RECEIVERS FOR CLASSROOM USE

Number of Television Receivers	Total Schools (81,000)	Total Pupils (44 million)
Total (all numbers)	100%	100%
None	25	19
1 to 4	52	51
5 to 9	13	15
10 or more	10	15

GLOSSARY

Amplifier: An electrical device that amplifies the voltage, current, or power of an electrical signal.

Cable television, cable TV, CATV: A system for distributing audiovisual information via coaxial cable. Includes its signal receiving, amplifying, and controlling equipment and signal origination equipment. The system is used under franchise granted by a governmental body.

Channel: A frequency band providing a single path for transmitting electrical signals, usually in distinction from other parallel paths.

Closed-circuit television, closed-circuit TV, CCTV: A system for distributing audiovisual information via coaxial cable. Includes its signal receiving, amplifying, and controlling equipment and signal origination equipment. The system is under control of the user (see Dedicated system).

Coaxial cable: The most commonly used means of signal transportation for cable TV, consisting of a cylindrical outer conductor surrounding a central conductor.

Common carrier: Microwave service available at published tariff rates approved by the FCC.

Computer assisted instruction, CAI: The use of a computer in a tutorial mode. The computer adapts the presentation of instructional material on the basis of inputs from the student.

Dedicated system: A transmission system operated entirely for a single user or a small group of users. The user determines content, scheduling, points of reception, and other matters having to do with program transmissions.

Digital data: Data and information that can be represented by a set of discrete items such as the integers.

Drop: The connection between the cable TV system and the subscriber's TV set.

Educational television, ETV: All noncommercial television intended for general use. The term includes public affairs, cultural, educational, entertainment, or other programming. The medium of broadcast television is generally used. The term includes ITV and PTV.

Film chain: A device for showing sound, motion pictures on TV.

Head-end: The electronic equipment located at the start of a cable system, usually including antennas, preamplifiers, frequency converters, demodulators, modulators, and related equipment. May also include the antenna tower and building housing the above.

Independent random access: Access to an item of instructional materials (including audiovisual materials) at random times and independent of the use of the item by another person.

In-school project: A project for transmitting television programming to the facilities of an education agency.

Instructional Television Fixed Service, ITFS: A service operated by an education organization to transmit audiovisual information to one or a few fixed receiving locations. Directional transmission is provided at frequencies higher than those used for broadcast television.

Instructional television, ITV: Television programming that has as its purpose the production, origination, and distribution of instructional content for people to learn. It is closely related to the work of organized formal education agencies.

Local production: Production of television programming by a cable TV operator, closed-circuit TV operator, ITFS operator, or other local agency, rather than by television networks or national agencies.

Microwave: Applies to transmission at frequencies well above the normal television frequencies.

Narrow-band: A range of frequencies less than 3000 cycles from lowest to highest.

Out-of-school project: A project for transmitting television programming to the home or other facilities that are not part of an education agency.

Penetration: The percent of total homes that have one or more television sets connected to a cable TV system, or being connected to a cable TV system, at the time of a survey.

Plasma panel: A computer terminal device for visual display. It consists of bubbles of gas that can be made to glow on and off by electric current.

Preferential treatment: The provision of cable system facilities or services to a subscriber free or at a lower cost than that charged the general subscriber.

Public television, PTV: ETV programming intended for the general community, as opposed to ITV.

Random access: Access to an item of instructional materials (including audiovisual materials) at random, as distinct from scheduled, times. Access may depend on whether the item is already in use by another person, however.

Trunk: The main cable in a cable system.

Two-way capability: An ability to transmit signals in both directions. Return signals may be digital, audio, or audiovisual.

Utilization: The use of televised programming for its intended (usually educational or instructional) purpose.

Utilization activities: Activities that abet utilization.

Video: Relating to or used in the transmission, reception, or recording of audiovisual information. Usually considered to be signals in the range of normal television frequencies.

Videotape: Magnetic tape carrying audiovisual information.

BIBLIOGRAPHY

Anaheim School District, Anaheim, California, Department of Instructional Media, "Teaching with Television," no date.

Baer, Walter S., *Cable Television: A Handbook for Decisionmaking,* Crane, Russak & Co., Inc., New York, 1974.

——, *Interactive Television: Prospects for Two-Way Services on Cable,* The Rand Corporation, R-888-MF, November 1971.

——, and R. E. Park, "Financial Projections for the Dayton Metropolitan Area," Paper 2 in Johnson et al., *Cable Communications in the Dayton Miami Valley: Basic Report,* The Rand Corporation, R-943-KF/FF, January 1972.

Belt, Forest H., "Education in Kentucky—By Television," *BM/E (Broadcast Management and Engineering),* October 1971.

Billowes, C. A., "On-Demand Educational Television Program Retrieval System for Schools," *Proceedings of the IEEE,* Vol. 59, No. 6, June 1971.

Bloom, B. S. (ed.), *Taxonomy of Educational Objectives. Handbook I: Cognitive Domain,* David McKay Company, Inc., New York, 1956.

Bogatz, Gerry Ann, and Samuel Ball, *The Second Year of Sesame Street: A Continuing Evaluation,* Educational Testing Service, Princeton, New Jersey, November 1971.

Bretz, R., "Closed-Circuit ITV Logistics—Comparing Hagerstown, Anaheim, and Santa Ana," *Journal of the NAEB,* July-August 1965.

——, "The Potential Uses of Cable in Education and Training," Paper 7 in Johnson et al., *Cable Communications in the Dayton Miami Valley: Basic Report,* The Rand Corporation, R-943-KF/FF, January 1972.

——, *The Selection of Appropriate Communication Media for Instruction: A Guide for Designers of Air Force Technical Training Programs,* The Rand Corporation, R-601-PR, February 1971.

——, *A Taxonomy of Communication Media,* Educational Technology Publishers, Englewood Cliffs, New Jersey, 1970.

——, *Three Models for Home-Based Instructional Systems Using Television,* The Rand Corporation, R-1089-USOE/MF, October 1972.

——, *Will My Visual Be Visible?,* The Rand Corporation, P-4919, October 1972.

The Carnegie Commission on Educational Television, *Public Television: A Program for Action,* Bantam Books, New York, January 1967.

The Carnegie Commission on Higher Education, *The Fourth Revolution—Instructional Technology in Higher Education,* McGraw-Hill Book Company, New York, June 1972.

Carpenter, C. R., "Instructional Film Research—A Brief Review," *British Journal of Educational Technology,* Vol. 2, No. 3, October 1971.

Carpenter, Polly, A. W. Chalfant, and G. R. Hall, *Case Studies in Educational Performance Contracting: 3. Texarkana, Arkansas, Liberty-Eylau, Texas,* The Rand Corporation, R-900/3-HEW, December 1971.

——, and S. A. Haggart, *Analysis of Educational Programs within a Program Budgeting System,* The Rand Corporation, P-4195, September 1969.

——, and G. R. Hall, *Case Studies in Educational Performance Contracting: Conclusions and Implications,* The Rand Corporation, R-900/1-HEW, December 1971.

——, and M. L. Rapp, *Testing in Innovative Programs,* The Rand Corporation, P-4787, March 1972.

——, et al., *Analyzing the Use of Technology To Upgrade Education in a Developing Country,* The Rand Corporation, RM-6179-RC, March 1970.

Chu, G. C., and W. Schramm, *Learning from Television: What the Research Says,* Stanford University, Stanford, California, 1967.

Connell, D. D., "How Sesame Street Raised and Spent Eight Million Dollars," *Dividend,* Graduate School of Business Administration, University of Michigan, Winter 1971.

Cooney, J. G., *The Potential Uses of Television in Preschool Education,* New York, October 1966.

——, *Television for Preschool Children: A Proposal,* February 19, 1968.

Dambrot, Faye, "General Psychology over Closed-Circuit Television," *AV Communication Review,* Vol. 20, No. 2, Summer 1972.

Educationql Product Report, No. 31, Instructional Television Fixed Service, Educational Products Information Exchange Institute, New York, 1971.

Educational Product Report, No. 36, Dial Access Systems and Alternatives, Educational Products Information Exchange Institute, New York, 1971.

Efron, Edith, "Peter Chelkowski, Ph.D., I Love You," *TV Guide,* November 21, 1970.

Erickson, C. G., and H. M. Chausow, *Chicago's TV College, Final Report of a Three Year Experiment,* Chicago City Junior College, Chicago, August 1960.

Etzioni, Amitai, "Human Beings Are Not Very Easy To Change After All," *The Saturday Review,* Vol. 55, No. 23, June 3, 1972.

"Experts Praise AEL's Home-Oriented Preschool Program," *Education Daily,* Vol. 6, No. 68, April 9, 1973.

Fact Book—Television Instruction, Indiana University Foundation, National Instructional Television Center, Bloomington, Indiana, 1972.

Farquhar, J. A., et al., *Applications of Advanced Technology to Undergraduate Medical Education,* The Rand Corporation, RM-6180-NLM, April 1970.

Federal Communications Commission, "Cable Television Report and Order," *Federal Register,* Vol. 37, No. 30, Part II, February 12, 1972.

——, "Memorandum Opinion and Order on Reconsideration of the Television Report and Order," *Federal Register,* Vol. 37, No. 136, Part II, July 14, 1972.

Feldman, N. E., *Cable Television: Opportunities and Problems in Local Program Origination,* The Rand Corporation, R-570-FF, September 1970.

——, "System Designs for the Dayton Metropolitan Areas," Paper 1 in Johnson et al., *Cable Communications in the Dayton Miami Valley: Basic Report,* The Rand Corporation, R-943-KF/FF, January 1972.

Fink, Alan, "Public Education's Stake in the Proposed Community Antenna System Franchise," *Schools and Cable Television,* Division of Educational Technology, National Education Association, Washington, D.C., 1971.

Garfunkel, Frank, "Sesame Street—An Educational Dead End?," *Bostonia,* Boston University Alumni Magazine, Winter 1970.

Haggart, S. A., *Program Cost Analysis in Educational Planning,* The Rand Corporation, P-4744, December 1971.

——, G. C. Sumner, and J. R. Harsh, *A Guide to Educational Performance Contracting: Technical Appendix,* The Rand Corporation, R-955/2-HEW, March 1972.

——, et al., *Program Budgeting for School District Planning,* Educational Technology Publications, Englewood Cliffs, New Jersey, 1972.

Hall, G. R., et al., *A Guide to Educational Performance Contracting,* The Rand Corporation, R-955/1-HEW, March 1972.

Hammond, A. L., "Computer-Assisted Instruction: Two Major Demonstrations," *Science,* Vol. 176, No. 4039, June 9, 1972.

Hill, Roger W., Jr., "Educational Considerations of CATV Cablecasting and Telecommunications," *Educational/Instructional Broadcasting,* Vol. 2, No. 9, November 1969.

"Interview with Dr. Donald G. Emery, Director National Reading Center," *Reading Newsreport,* Vol. VI, No. 5, March 1972.

Johnson, L. L., *Cable Television and Higher Education: Two Contrasting Examples,* The Rand Corporation, R-828-MF, September 1971.

——, *The Future of Cable Television: Some Problems of Federal Regulation,* The Rand Corporation, RM-6199-FF, January 1970.

——, et al., *Cable Communications in the Dayton Miami Valley: Basic Report,* The Rand Corporation, R-943-KF/FF, January 1972.

Jones, G. R., *The Jones Dictionary of CATV Terminology,* Johnson Publishing Co., Boulder, Colorado, 1971.

Kenney, B. L., and F. W. Norwood, "CATV: Visual Library Service," *American Libraries,* July-August 1971, pp. 723-726.

Klaus, D. J., *Instructional Innovation and Individualization,* American Institutes for Research, Pittsburgh, Pennsylvania, 1969.

Kratochvil, Daniel W., *Sesame Street: Developed by Children's Television Workshop,* Product Development Report No. 10, American Institutes for Research in the Behavioral Sciences, Palo Alto, California, December 1971.

Lee, J. A., *Test Pattern,* University of Toronto Press, Toronto, 1971.

Levien, R. E., *The Emerging Technology: Instructional Use of the Computer in Higher Education,* McGraw-Hill Book Company, New York, 1972.

Maclure, S., "England's Open University Revolution at Milton Keynes," *Change,* Vol. 3, No. 3, March-April 1971.

Martin-Vegue, Charles A., Jr., et al., "Technical and Economic Factors in University Instructional Television Systems," *Proceedings of the IEEE,* Vol. 59, No. 6, June 1971.

McCarty, H. R., *Proposal for Development of Electronic Communication System for the San Diego Region,* Department of Education, San Diego County, California, January 17, 1972.

McCombs, M., "Chicago's Television College," *New Educational Media in Action: Case Studies for Planners,* Vol. 2, United Nations Educational, Scientific and Cultural Organization, Holland-Breumelhof N.V., Amsterdam, 1967.

McLaughlin, G. H., et al., *Educational Television on Demand,* Occasional Papers No. 11, The Ontario Institute for Studies in Education, Toronto, 1972.

Molenda, M. H., "The Educational Implications of Cable Television (CATV) and Video Cassettes: An Annotated Bibliography," *Audiovisual Instruction,* Vol. 17, No. 4, April 1972, pp. 42-59.

National Education Association, *Cable Television: Franchise Provisions for Schools,* Instruction and Professional Development, NEA, Washington, D.C., February 1973.

——, *Schools and Cable Television,* Division of Educational Technology, NEA, Washington, D.C., 1971.

——, *A Survey of Instructional Closed-Circuit Television, 1967,* NEA, Washington, D.C., 1967.

National Instructional Television Center, *Newsletter,* Vol. IV, No. 1, September/October 1972.

——, *Ripples,* Fact Sheet No. 64.

1972 Recorded Instruction for Television, University of Nebraska, Great Plains National Instructional Television Library, Lincoln, Nebraska, 1972.

Noel, E. S., and G. M. Helmke, *Instructional Television in California,* California State Department of Education, Sacramento, California, 1968.

Ohio Bureau of Employment Services, *Occupational Survey, 1970-71, Akron Metropolitan Area,* April 1970.

Park, R. E., *Prospects for Cable in the 100 Largest Television Markets,* The Rand Corporation, R-875-MF, October 1971.

Price, M., and J. Wicklein, *Cable Television: A Guide for Citizen Action,* Pilgrim Press, Philadelphia, Pennsylvania, 1972.

Quick, John, and Herbert Wolf, *Small-Studio Video Tape Production,* Addison-Wesley Publishing Co., Reading, Massachusetts, 1972.

Rapp, M. L., *Evaluation as Feedback in the Program Development Cycle,* The Rand Corporation, P-4066, April 1969.

Reeves, B. F., *The First Year of Sesame Street: The Formative Research,* Children's Television Workshop, New York, December 1970.

Reuben, Gabriel H., "Using Cable Television To Involve Parents," *Educational Television,* III, January 1970.

Robinson, W. R., *Analysis of Data from the IRTV Overload Study,* Bell Telephone Research, Ltd., Ottawa, Canada, May 25, 1971.

Rockman, S., *One Week of Educational Television, No. 6, March 9-15, 1970,* National Instructional Television Center, Bloomington, Indiana, 1971.

Rubinstein, E. A., et al., *Television and Social Behavior, Vol. IV, Television in Day-to-Day Life: Patterns of Use,* U.S. Government Printing Office, Washington, D.C., 1972.

Samuels, B., *The First Year of Sesame Street: A Summary of Audience Surveys,* Children's Television Workshop, New York, December 1970.

Schmidbauer, M. (Project Manager), *Multimedia System in Adult Education,* Internationales Zentralinstitut fur das Jugend- und Bildungsfernsehen, Munich, 1971.

Schramm, Wilbur, et al., *The New Media: Memo to Educational Planners,* United Nations Educational, Scientific and Cultural Organization, Holland-Breumel-hof N.V., Amsterdam, 1967.

Shafer, Jon, *A Cable TV Guide for Educators: Cable Communications Comes to the Twin Cities Area,* Educational Research and Development Council of the Twin Cities Metropolitan Area, Inc., University of Minnesota, St. Paul, January 1972.

Shanks, Robert E., "The Anaheim Approach to Closed-Circuit Television," *A Guide to Instructional Television,* ed. Robert M. Diamond, McGraw-Hill Book Company, New York, 1964.

Sivatko, J. R., "Newest Teaching Method Provided by Educasting," International Correspondence Schools, Scranton, Pennsylvania, 1964 (mimeographed press release).

TAGER, *Annual Report, 1969-70,* The Association for Graduate Education and Research of North Texas, October 1970.

"Teacherless Classes," *Nation's Schools,* Vol. 89, No. 6, June 1972.

Television Factbook, Services Volume, No. 41, Television Digest, Inc., Washington, D.C., 1971-1972 edition.

Tickton, S. G. (ed.), *To Improve Learning: An Evaluation of Instructional Technology,* 2 vols., R. R. Bowker Company, New York, 1970, 1971.

U.S. Department of Commerce, Bureau of the Census, *Educational Attainment in 30 Selected Standard Metropolitan Statistical Areas: 1970,* P-20, No. 227, November 1971.

——, ——, *Employment Profiles of Selected Low-Income Areas: New York, N.Y.-Puerto Rican Population of Survey Areas,* PHC(3)-3, January 1972.

U.S. Department of Health, Education, and Welfare, Office of Education, *Career Education Research and Development Program,* Washington, D.C., June 15, 1972.

——, ——, National Center for Educational Statistics, *Bulletin,* No. 7, February 9, 1971.

U.S. Department of Labor, Office of Economic Opportunity, *An Experiment in Performance Contracting: Summary of Preliminary Results,* OEO Pamphlet 3400-5, February 1972.

U.S. Government Films: A Catalog of Motion Pictures and Filmstrips for Rent and Sale by the National Audiovisual Center, General Services Administration, National Archives and Records Service, Washington, D.C., Publ. 72-17, 1971 Supplement.

Wade, S., "Hagerstown: A Pioneer in Closed-Circuit Televised Instruction," *New Educational Media in Action: Case Studies for Planners,* Vol. 1, UNESCO, International Institute for Education Planning, Amsterdam, 1967.

Wagner, R. V., et al., *A Study of Systemic Resistances to Utilization of ITV in Public School Systems, Volume II, Case Studies,* American University, Washington, D.C., February 1969 (ED 030 012).

Wigren, H. E., "The NEA's Position on Cable Television," *Schools and Cable Television,* Division of Educational Technology, National Education Association, Washington, D.C., 1971.

Wilson, S. W., "Interactive Lectures," *Technology Review,* January 1972.

Part IV

A Guide for Education Planners
Polly Carpenter-Huffman

I. INTRODUCTION

Purpose of the Guide

The past few years have seen a rapid growth of cable television throughout the United States, both in terms of the number of systems and in terms of the development of cable technology. With this growth has come an increasing awareness of the potential of cable TV for providing more than improved quality of signal in fringe areas for broadcast TV. Cable TV now has the capacity to provide better, more, and new telecommunication services. Recognizing this potential, the Federal Communications Commission (FCC) recently issued a set of regulations for the installation and operation of cable TV systems.[1] These rulings have given rise to increased interest in the use of cable TV for public services.

Since the early 1950s, television has offered the education community means for solving or alleviating a number of its problems. A few individuals and school systems have recognized these possibilities and taken advantage of them; but they have been too few. By and large, the education community has ignored television except as a superficial token of its support of innovation.

Because of this history, the FCC has adopted a wait-and-see attitude toward the education community's requests for a large measure of "preferential treatment"[2] in the operation of new cable TV systems. This is reflected in the FCC's recent set of regulations which avoid locking cable operators into arrangements preferential to schools—arrangements that the education community might let lie fallow. It is our belief that the FCC and a number of cable TV operators would welcome greater use of television in education. But this use must be significant in terms of the contribution it makes to the improvement of education; mere acquisition of a few TV sets will not turn the trick.

The purpose of this Guide is to help education planners develop viable plans that will demonstrate the commitment of the education community to making significant use of cable TV. Viable plans will go far toward ensuring that the poten-

[1] FCC, "Cable Television Report and Order." See also FCC, "Memorandum Opinion and Order on Reconsideration of the Television Report and Order."

[2] Preferential treatment would provide the education community with cable system facilities or services free or at lower rates than the general subscriber pays.

tial usefulness of cable TV to education will be realized before commercial interests have usurped the cable system. Viable plans will also support special showings to the FCC for the preferential treatment that may be needed in some cases. The FCC has invited educators to present such showings.

Unless cable TV is to be only another frill, its use should be integrated with other educational functions. This requires careful planning, development, and evaluation—processes this Guide treats in detail. It is hoped these guidelines will encourage the development of genuinely useful applications and will discourage the proliferation of ill-conceived projects that turn out to use television for little more than custodial services for the teacher's convenience.

Many of the skills needed for planning, development, operation, and evaluation of the project[3] would be needed for implementing and managing any new education activity and may be available within an existing school system. Others, particularly those having to do with the technical aspects of the cable TV system and with obtaining and evaluating television programming, may be more difficult to find. The Guide includes suggested sources of special skills that may be lacking in the local area.

It is particularly difficult to obtain television programming of acceptable quality that is appropriate to the user's needs. Procedures for obtaining and judging prerecorded materials are outlined. The alternative, to produce original programming, is discussed briefly, especially from the point of view of the costs involved and possible arrangements for reducing the impact of cost on a single education agency. Other publications[4] go into the details of small-scale production of TV programming.

Basis of the Guide

Our "dos" and "don'ts" are based on the experiences of schools and other education institutions with television, the recommendations of leading groups in education or television or both, and our own backgrounds in planning, management, and evaluation of education projects and in communication technology and policy.[5] Some of the more significant projects using television in education are discussed in the preceeding part. A selected bibliography is supplied to provide richer detail as the reader desires.

Plan of the Guide

There are seven chapters. The first two set the stage by defining the problem and discussing special aspects of cable TV as a communication medium. Chapters

---

[3] A note on terminology. We use *project* rather than *program* to avoid confusion with a *televised program*. *Project* has a short-lived connotation which we do not intend, but there seems to be no better word.

[4] See, for example, Quick and Wolf, *Small-Studio Video Tape Production.*

[5] Some Rand studies in the fields of communication and education are included in the Bibliography.

3 through 7 set forth the guidelines for implementing an education project using cable TV. After an introductory chapter describing the implementation process in general, Chap. 4 presents the details of defining and planning an in-school project (a project that transmits programs to facilities of an existing education institution). Chapter 4 is the most technical and is intended to be of substantial assistance to the project manager. The implementation of an out-of-school project (that is, one that does not transmit to the facilities of an existing institution) is different from that of an in-school project. These differences are noted and discussed in Chap. 5. The development and implementation phases are dealt with in Chap. 6. Finally, because of its importance to project success, evaluation is treated separately in Chap. 7.

II. CABLE TELEVISION AS A MEANS OF DISTRIBUTION

We start by comparing cable TV with other means of distributing audiovisual information that are important to education. Comparisons include salient characteristics of distribution systems and the effects that changes in the way they are used have on costs. Much of the usefulness of a cable TV system may hinge on provisions written into the franchise awarded to the cable operator. This is why we talk about the franchising process for a good part of this chapter.

Comparing Means of Distribution

Cable TV is only one of several ways to distribute audiovisual information. Its essential features include a headend[1] for amplifying and processing signals for cable transmission, coaxial cables to carry the signals to subscribers, terminal equipment connected to subscribers' TV receivers, and a group of antennas for receiving broadcast television signals for retransmission over the cable. Studios for local program production can also be connected to the headend. Although the cable may be buried underground, it is usually strung from utility poles; amplifiers to boost signal strength are positioned at intervals along the cables.

Because miles of coaxial cable and electronic equipment must be installed, initial investments in cable TV are relatively high. For example, of the $7.5 million estimated for capital investment for a Dayton cable TV system, $6.5 million (over 85 percent) will be required for buying and installing the cable distribution plant.[2]

Other transmission media of importance to education are the familiar broadcast TV, closed-circuit TV (CCTV), and instructional television fixed service (ITFS). The primary physical difference between broadcast TV and cable TV is that the television signals are transmitted over cable rather than broadcast. Cable TV and CCTV are physically identical. They are *closed* in the sense that signals they originate cannot be picked up off the air by anyone who has a TV receiver. The important difference is that a cable TV system accommodates a number of independent users

---

[1] See the Glossary for definitions of technical terms.
[2] Baer and Park, "Financial Projections for the Dayton Metropolitan Area."

(including a school system, for example), but in a CCTV system a single user or a small group of users control program content and scheduling and direct program usage. Although someone else, such as the telephone company, may own the physical plant, the user directs its utilization. Such systems are termed *dedicated* systems. Frequently, closed-circuit systems are installed in a single building or a complex of buildings, such as a university campus or industrial plant.

A number of education institutions have adopted or are planning to use ITFS for providing TV transmission. ITFS uses much higher frequencies and much lower power than broadcast TV, and antennas and receivers required to pick up the signal are expensive. Each qualified applicant for an ITFS license may obtain up to four channels within the spectrum space allotted. This space may be used not only for video but for both voice and data transmission. In addition, part of the spectrum space may be used for response in the form of video, voice, or data. ITFS systems are also dedicated systems.

In Table 1, some features of these four means of distribution are compared. Note that although the channel capacity of broadcast television is moderately high, only one or two of these channels can be set aside for educational broadcasting in a specific area. On the other hand, all of the channels on either ITFS or closed-circuit systems are at the service of the person who controls the system. In a cable TV system, the number of channels that can be leased to the education community (beyond the free channel mandated by the FCC) can be very large if the system has the capacity needed. The best way to ensure this is to make arrangements during the franchise negotiations and before the cables are installed.

Using more than one trunk from the headend is another way to increase the channel capacity of a CCTV or cable TV system without adding cables. The number of channels available within a single system can be multiplied by a factor of two to four; most systems have two trunks. Of course, only the programming on a single trunk is available to an individual receiver.

Figure 1 shows how the audience for instructional television (ITV) programming was distributed by grade level in 1970.[3] Note that the bulk of broadcast ITV programming was directed to the lower grades; less than 30 percent was provided above the junior high level. One explanation for this phenomenon is that the departmentalization of subject matter above the elementary grades makes scheduling of the single channel an extremely complicated task. Additional channels would greatly alleviate this difficulty.

Data are also given for ITFS and CCTV systems to demonstrate the difference in usage of these two dedicated media. ITFS shows a similar pattern of usage to that of broadcast ITV, whereas CCTV does not. Instead, CCTV seems more attractive to institutions of higher education than is ITFS. This is probably because ITFS becomes competitive in cost with CCTV when an area larger than a single campus is to be served.

Despite the fact that both ITFS and CCTV make it possible for schools to use multiple channels, relatively few of the schools that have either type of system take advantage of this opportunity. For example, during a week in 1970, of those schools that had CCTV systems, 49 percent used only one channel. Channel use for ITFS

[3] Rockman, *One Week of Educational Television.*

Table 1

COMPARISON OF TV TRANSMISSION MEDIA

Feature	Transmission Medium			
	Broadcast	Cable	Closed-circuit	ITFS
Transmission type	Broadcast	Coaxial cable	Coaxial cable	Broadcast
Maximum channel capacity	11-13	12-35 per cable[a]	12-35 per cable[a]	4
Channels for education	1-2	1 free[b]	12-35 per cable	4
Two-way capability	No	Possible	Possible	Yes
Privacy	Can be provided	Can be provided	Inherent	Inherent
Nominal radius of coverage	20-140 miles (depends on topography)	5 miles	5 miles	8-25 miles (depends on topography)
FCC regulations	Issues station license	Oversees franchise provisions	None	Issues system license

[a] Multiple trunks can increase these numbers by factors of two to four.

[b] More can be leased. The maximum number depends on negotiations prior to franchising.

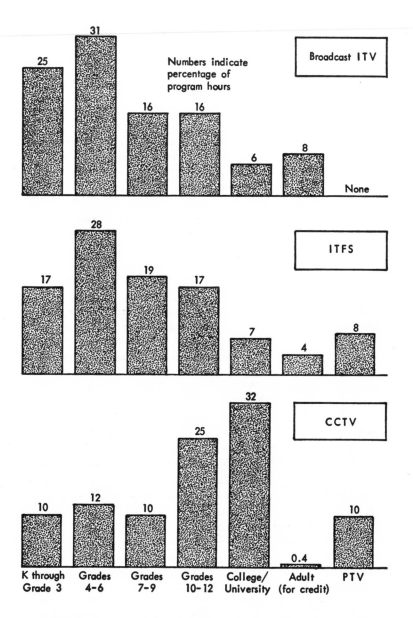

Fig. 1—Distribution of ITV programming by intended audience, March 1970
(from Rockman, *One Week of Educational Television*)

systems was somewhat better, with 77 percent using more than one channel (usually two).

Cable TV has a capacity for two-way communication, but so does CCTV, and a number of ITFS systems are already using voice talk-back. Talk-back encourages the student to respond during program transmission and thereby, in contrast with broadcast TV, not only increases his involvement in the program but also can make the program more appropriate to his needs.

Two-way capability may take one of several forms. Probably the most common will be narrow-band channels for return of digital data. Digital data return could be useful in a wide variety of applications, including programmed instruction. Tens of thousands of individual students could be accommodated in the space occupied by a single video channel. In contrast, on the order of only six hundred channels of audio signals could be accommodated because such signals require a far larger bandwidth than do digital data signals. Audiovisual information, of course, requires the greatest bandwidth of all; the number of return channels of audiovisual information that can be provided will be somewhat less than the 12 to 35 available downstream on a single cable.

The cost to the user for the several forms of return capability will also vary. Baer has estimated that the annual cost to the operator per two-way subscriber for narrow-band services will run between $100 and $500, depending on the services supplied. About $50 of this is for the one-way cable service; the remainder is the cost of adding two-way service. A shared voice channel would add some $100 to the cost of the subscriber's terminal; a shared video channel would add between $700 and $1500 to the cost of his terminal.[4]

Privacy is inherent to both CCTV and ITFS systems because these are dedicated systems. Privacy can be obtained on both broadcast and cable systems by using scrambled signals.[5]

Why is privacy desirable? First, few people can act naturally in front of an unfamiliar audience. Exposing students and teachers to the glare of the public eye would undoubtedly inhibit all but the most extroverted. Second, the public is notoriously intolerant on hot issues. How can we expect the schools to deal with anything but the blandest of subjects if the public is sitting in? Finally, some subjects are not fit for just anyone to watch. Imagine broadcasting some of the procedures that doctors or nurses must learn.

ITFS and broadcast TV can cover a larger geographic area than cable or closed-circuit TV. If mountains are within range, the radius of coverage of the broadcast systems can be on the order of a hundred miles. ITFS lends itself to the use of several schools or agencies located through a relatively large region. CCTV frequently has been used within a single building or complex of interconnected buildings, such as a campus, as suggested in Fig. 1. To serve areas larger than 5 miles or so in radius, interconnected cable systems must be used. Thus, as of 1971, 65 ITFS systems were on the air serving 119 schools and an additional 14 were in the construction phase,[6]

---

[4] Baer, *Interactive Television.*

[5] It has been proposed that nonstandard channels with special converters would provide privacy on cable TV systems. This is true in principle, but converters are inexpensive and readily obtainable. Thus, the use of converters will not *ensure* privacy.

[6] *Educational Product Report, No. 31.*

whereas nearly 7000 schools were using CCTV systems in 1970. Nearly 54,000 schools made some use of broadcast TV.[7]

The FCC licenses both broadcast TV stations and ITFS systems and establishes rules governing the franchises of cable TV systems. CCTV systems, of course, are unregulated. Because ITFS systems are private, in a sense, the FCC is relatively relaxed about their operations. The strongest restrictions are put on the operations of broadcast TV stations; what the FCC rulings on cable TV franchises will mean to the education community remains to be seen.

Table 1 shows that CCTV offers the greatest flexibility for educational use, if the geographic coverage needed is small. Why, then, would you be interested in cable TV? The interest springs from the hope to have a system like CCTV for a much lower cost. The trouble with CCTV is that a single institution (or a coalition of a few institutions) pays for the system because it is dedicated to the use of the institution. In contrast, the cable operator spreads his costs over thousands of users and can still make money even if he provides free channels, drops, and intra-school connections. Thus, by hooking onto an existing cable system, education agencies can greatly reduce the costs to interconnect school buildings for receiving instructional programming. If you need more capacity than will be available by leasing channels, a special *extra* cable system of dedicated lines can be run along the same routes as the basic system. If installed at the same time as the basic system, its cost would be on the order of half the cost of a comparable CCTV system.

Of course, if you *require* multiple channels and your receivers are within a small area, such as a single building or campus, CCTV can still be attractive. This is especially true if the cost of teaching a single student is high, and it explains why many medical schools use CCTV.

But again there is the limitation on geographic coverage. If schools are separated by more than a few miles, you may need an ITFS system or highly directional microwave transmission to reach them all. Like a CCTV system, an ITFS system will cost more than cable because it is a dedicated system.[8] In addition, four channels, at most, will be available. Again, if the cost of teaching a single student is high, an ITFS system could be attractive.

The biggest disadvantage of ITFS for a large school system is the cost of the special receiving equipment that you must add at each reception point—a cost that denies you the economies of scale available in other systems. The best solution for a large, geographically extensive system would be a combination of systems, with ITFS or microwave to span the longer distances and cable within smaller regions.

The interconnected system might still offer only four channels because of the limitations on ITFS. If the system serves a consortium of education agencies, however, the FCC may allow more channels. For example, the FCC approved the application for 16 channels by ETAMC, a consortium in Cleveland (discussed more fully below). Given additional money for development of ITFS system equipment and the

---

[7] HEW, Office of Education, National Center for Educational Statistics, *Bulletin*, No. 7, February 9, 1971.

[8] Costs are also high because producers of ITFS systems cannot make the concessions that can be made by cable TV operators because of the large subscriber base.

A cost comparison of nominally equivalent cable TV, CCTV, and ITFS systems is given in App. A. Costs are highly contingent on specifics of geographic coverage, number of channels, number of schools, and other matters.

approval of the FCC, 20 or more channels might be fitted into the ITFS spectrum space.

Ultimately, most of the problems of limited channels should be solved by the use of media that are as easy to access as the printed page. At present, videotape players are unreliable and incompatible among manufacturers, film projectors are awkward to operate, and videotapes and films are too expensive for every classroom to have a full set. Audiotape cassettes and players are becoming as portable and reliable as, say, printed materials, but they are still expensive; also it is hard to locate specific material on a tape. Film and videotape cassettes are now within the economic reach of some education and training institutions and will certainly be used more widely in the future. For example, the U.S. Army plans to have 6000 to 10,000 video cassette players before mid-1973.[9] Wide availability of inexpensive videotape and film cassettes and players will undoubtedly alter some of the roles of television. For example, you could use cable TV during off-hours to send programming to videotape recorders; recorded programs could then be used at the teacher's discretion.

Systems for distribution of television signals may consist of combinations of cable TV, ITFS, CCTV, and standard broadcast transmissions. Many existing combinations have grown together with the passage of time during which a single source enlarged the geographical area of its operations. In planning to use television for your school or training institution, you should consider what transmission means are now available and how cable TV might supplement or supplant them. Other means of transmission might be useful adjuncts in some areas or circumstances.

The Cable Television Franchise [10]

How well a cable TV system will meet your needs depends not only on your interest and initiative but also on arrangements that are made for cable service to your education community. These arrangements are usually (but not always) spelled out in the franchise under which the cable system will be operated.

The FCC has required that newly franchised operators in the 100 largest television markets provide one free channel for educational use during an experimental period of at least five years. Systems franchised before March 31, 1972, will have at most five years to comply with this requirement. You may be able to negotiate the lease of additional channels or to use facilities that are not needed full-time by the cable operator. The operator may even be eager to cooperate with you to increase his output of the locally originated programming, which he is required by law to supply if he has more than 3500 subscribers. And more than a single free channel may be made available if you can make a special showing of need and plans for use to the FCC.

The meaning of the FCC ruling on "preferential treatment of community groups" is unclear. Right now, the FCC seems to view 97 percent of the cable system revenues as a nominal minimum return to the cable operator. Any franchise provisions that would have the effect of giving the cable operator a lower return will probably be treated as potentially burdensome and as requiring a special showing.

[9] *The Video Publisher.* Vol. 2, No. 5, July 12, 1972.
[10] This section has been condensed from Chap. 4 of Part III of this book.

The FCC is much more likely to consider requests for preferential provisos favorably if you can support your need with a viable plan and can demonstrate the ability of the cable system to provide what you ask for. Thus, the importance of careful planning cannot be overemphasized. Only after people like yourself have argued a series of well-supported cases to the FCC will the rulings be clarified and will it be possible for the education community to take certain forms of preference for granted. You should view all of the possibilities for preferential treatment discussed here in this light.

Of course, preferential treatment that adds no financial burden to the cable operator can be provided without requiring special justification before the FCC.

FRANCHISE NEGOTIATION

The FCC requires that local franchise proceedings include a "public hearing affording due process." You should join representatives of the local school district, institutions of higher education, and other institutions concerned with preschool or adult education in these hearings. After the franchise is awarded, you should promote the establishment of an officially recognized agency that will protect the public's interest in the system.

ADMINISTERING THE EDUCATION CHANNEL

The regulations require that the education channel be made available by the operator free of charge, that advertising, lotteries, and obscene and indecent material be barred, and that a list of applicants for the channel be maintained for two years. Furthermore,

> Except on specific authorization, or with respect to the operation of the local government access channel, no local authority shall prescribe any other rules concerning the number or manner of operation of access channels. . . .[11]

There is a real danger that the cable operator may restrict the channel to a single school agency to simplify management or for some other reason. To ensure that the needs of the community or of special community groups will not be sacrificed in the cable TV operator's desire for profits, a separate entity comprising several users may be established to have jurisdiction over the education channel. The National Education Association has recommended the establishment of an advisory board to coordinate the use of the education channel.[12]

Joint agencies will also help ensure that the needs of all potential users are considered fairly. Ideally, the education channel will be allocated to those who can

---

[11] FCC, "Cable Television Report and Order."

[12] NEA, *Cable Television: Franchise Provisions for Schools.*

make best use of it in terms of need and cost. Criteria for selection of users should include

- The number in each potential audience.
- The importance of the potential use of the cable TV system.
- The extent to which the cable TV system provides services that cannot be provided better—if at all—in some other way.
- The cost (including the cost to the audience) of providing the educational service in some other way.

You will find it useful to partition potential users between daytime and evening hours. During daytime hours, the largest potential audiences are in schools or other service institutions or at home. For evening hours, consider programs for working adults or college students.

The strongest argument in favor of joint administration is that much more favorable financial arrangements can result for the support of local program production and other expensive items, such as computers and software repositories.

PREFERENTIAL FRANCHISE PROVISOS

According to the new FCC rules, the franchising authority must obtain FCC approval before requiring the lease of channels to the schools on a preferred basis or free.

> Because of the federal concern, local entities will not be permitted, absent a special showing, to require the channels be assigned for purposes other than those specified above. We stress again that we are entering into an experimental or developmental period. Thus, where the cable operator and franchising authority wish to experiment by providing additional channel capacity for such purposes as public, educational, and government access —on a free basis or at *reduced charges* —we will entertain petitions and consider the appropriateness of authorizing such experiments to gain further insight and to guide further courses of action.[13]

On the other hand, the FCC rules *require* cable operators to lease additional channels as needed. Therefore, if you need more than the free channel, your school may be able to lease them, perhaps at a preferential rate. For example, a charge could be made

> covering at least the incremental or out-of-pocket cost of supplying the channels, but... *less* than a proportionate share of the overhead of the cable operation in comparison with the charges for other services.... Such a pricing practice, in which some services bear much more of the overhead than others, is common in virtually all industries....[14]

As long as *all* users pay more than the cable operator's marginal cost of providing a leased channel, preferential rates are not a burden. At the least, the franchise

---

[13] Emphasis added. FCC, "Cable Television Report and Order."
[14] Baer et al., *Cable Television: A Handbook for Decisionmaking.*

should contain a guarantee that as cable leasing time becomes available the education community will have access to it.

According to a survey conducted by the National Cable Television Association in 1972, cable operators routinely give preferential treatment to the education community in a number of ways. For example, of the approximately 85 percent of the operators who provide regular service to schools, 70 percent provide free connection of schools to the cable system.

SCOPE[15] suggested (before the FCC ruling) that a franchise should include the provision

> That the grantee (CATV operator) will install without charge into each school building, public and private, college and library, receiving terminal and cable connections. . . .[16]

Twenty-five percent of the cable TV operators also reported that they made studio facilities or equipment available to the schools for production of school programming. Of these, 75 percent allowed schools to use the facilities or equipment without charge. If a large school population will use the cable system, it may be wise to include provisions in the franchise for production arrangements suitable to the school's needs.

Nothing in the FCC rules directly prohibits such preferential treatment. Yet, if additional channels are not to be made available routinely at reduced prices, the offering of basic and additional service at preferential rates may also be overruled by the FCC as a matter of consistency.

The FCC now requires that the "capacity" for nonvoice return communications be built into new cable systems in the major television markets. This means only that the system should eventually be able to provide return communications without "time-consuming and costly system rebuilding." Most operators will probably not install even the capacity for return transmissions at the outset. In a few instances, low-cost digital feedback links will be provided; these would be sufficient for such commercial applications as surveys, marketing services, and burglar alarm devices. Similar terminals could be used for instruction calling for selected response from the student, but you should not expect to use this capacity for 5 to 10 years unless your education agency will pay for the terminals.

The FCC has set minimum technical standards for broadcast television signals carried by cable TV systems but has set no standards at all for nonbroadcast services. You should consider the desirability of setting additional standards, which the franchisor may do if he will enforce them himself. At the least, the franchise should contain provisions that will require the incorporation of new standards as they are adopted.

In sum, you will want to ensure that the franchise agreement makes provisions for

- Access to additional channels for lease and leasing charges.
- Connection of school buildings to regular cable service and installation and monthly charges for that service.

[15] SCOPE: Suffolk County Organization for the Promotion of Education.

[16] Roger W. Hill, Jr. (SCOPE's director), "Educational Considerations of CATV Cablecasting and Telecommunications."

- Charges for making multiple connections within school buildings.
- Providing or sharing facilities and equipment for local production.
- An early date for operational two-way capability.
- Incorporation of new technical standards as they are set by governmental agencies.

ASSISTANCE IN FRANCHISE NEGOTIATIONS

Because many school systems and other education agencies lack personnel who are familiar with the technical and legal aspects of cable TV franchising, a number of agencies have formed to assist schools in taking advantage of this new technology. Some of these agencies are listed below:

PubliCable, National Education Association, Harold Wigren, 1201 Sixteenth Street, N.W., Washington, D.C. 20036. Telephone: (202) 833-4120. Active in assisting educators in using cable.

Joint Council on Educational Telecommunications, Frank W. Norwood, 1126 Sixteenth Street, N.W., Washington, D.C. 20036. Telephone: (202) 659-9740. Has information on cable developments in education.

Suffolk County Organization for the Promotion of Education, Roger W. Hill, Jr., Suffolk Educational Center, Stony Brook, New York 11790. Telephone: (516) 751-8500. Will advise on franchise negotiations for educational applications.

Office of Communication, United Church of Christ, 289 Park Avenue South, New York, New York 10010. Active in encouraging citizen participation in the communication field, particularly of minority groups. Will provide free literature on request.

III. OVERVIEW OF IMPLEMENTING A PROJECT USING CABLE TV

This chapter orients you to the next four chapters, each of which describes an aspect of the process of implementing an education project using cable TV. Here we discuss these aspects in general terms, and illustrate the time and resources required for implementation from the history of TV in education.

The four phases in implementing and operating an education project are

- Project definition and funding
- Planning
- Development
- Operation

Table 2 displays these four phases, along with the tasks that may be undertaken in each phase. Of course, if your project falls within funded activities, the fund-raising aspect of the first phase may be omitted. If you are contemplating a highly innovative project, most of these tasks will be necessary; if, however, you intend the project to supplement existing instruction, only the tasks denoted by bullets may be essential.

Two sets of tasks shown on the table are dealt with in separate chapters. These have to do with the cable TV system per se and with evaluation. Chapter 2 touched on tasks having to do specifically with the cable TV system; their precise content depends on the stage of implementation of the cable system and the franchising process. Evaluation is treated separately because it is crucial to successful implementation of innovative projects and because it interrelates with implementation and operation in complex ways. See Chap. 7.

The other tasks are in the mainstream of project implementation and operation and will be emphasized in the following three chapters. Here, each phase and each task will be described briefly. Before we begin, two points should be made: First, there are usually no neat dividing lines between the phases. For example, designing instruction should be continuous, with topics being refined on the basis of feedback from teachers and students. Similarly, the structure of the supporting organization should be designed while lines of authority and responsibility are being established. Second, tasks in one phase frequently overlap tasks in another, so that the total time

Table 2

STEPS IN IMPLEMENTING AN EDUCATION PROJECT USING CABLE TV[a]

Project Definition and Funding	Planning	Development	Operation
o Identify target audience o Establish general goals	Define target audience Specify project objectives		
	Design evaluation	Perform formative evaluation	Perform summative evaluation

Identify key resources	o Design project
	
	  TV programming
	
	  Concurrent instruction or counseling
	
	  Feedback from audience
	
	  Publicizing activities
	
	  Liaison with other groups
	
	o Estimate resource requirements
		Outline and integrate materials and programs
		
		o Schedule production or acquisition of materials and programs
		
		o Schedule distribution of materials and programs
		
		o Write and produce, or acquire, materials and programs
		
		Prepare for concurrent instruction or counseling
		
		o Arrange for feedback from audience
		
		Begin publicizing project
		
		Establish liaisons with other groups
			Enroll students
			
			o Produce or acquire and distribute materials and program
			
			Provide concurrent instruction or counseling
			
			o Service audience through feedback mechanisms
			
			o Monitor project performance
			
			Continue publicizing project
			
			Maintain liaison with other groups

o Determine role in franchise negotiations	o Establish liaison with cable TV operator Determine requirements for cable TV system support	o Maintain liaison with cable TV operator o Connect users to cable TV system	o Maintain liaison with cable TV operator
Identify funding sources Submit applications for funding	Define organizational structure	Obtain and train staff Set up organization	

[a]Bullets indicate minimum tasks.

required for implementation is rarely the sum of the times required for each phase. Thus, Table 2 shows a logical sequence that does not represent any specific project.

Project Definition and Funding

The project definition and funding phase can take from a few weeks to several years. The time required depends on how innovative the envisaged project is and how strongly it is opposed. In general, the more closely the project conforms with current classroom practice, the less effort will be required. For example, a decision to use educational programs offered by the local cable TV system to supplement classroom teaching may entail virtually no funding or project definition efforts. The funding phase for the Chicago TV Junior College, which essentially transferred the basic college curriculum to broadcast television, took only a few weeks.[1] The proposal for *Sesame Street,* on the other hand, drew on a study by Joan Cooney for the Carnegie Corporation that spanned nearly a year.[2] And the funding and project definition phase for England's Open University stretched over three to four years of intensive political maneuvering.[3]

Another decision is whether the project will be basic to the school's operations or will merely supplement them. If you have decided to use TV to expand the accessibility of education, decrease costs, or improve quality, the uses will probably be basic to your operations. If you are thinking only about added services, the uses are supplementary. The distinction is essential, because basic uses can require considerable effort to integrate the project with other school activities; supplementary uses rarely, if ever, levy such requirements. Most current uses of TV in schools fall into the supplementary class and usually involve little more than a few classrooms equipped with TV sets, a small amount of interior wiring, and a rooftop antenna to pick up broadcast programs. Such uses really don't deserve to be called "projects"; in most cases they have involved no discernible planning or development.

The sheer scope of the project and other considerations can also introduce delays. For example, the Kentucky ETV network provides supplementary programming for schools throughout the state. It

> was born September 1968, but gestation began 10 years earlier. Marrying education to technology became the self-appointed task of O. Leonard Press, who in 1958 was head of the radio-TV films department at the University of Kentucky (UK).... In 1960 the Kentucky general assembly agreed to a feasibility study. In 1962 it created the Kentucky Authority for Educational Television. O. Leonard Press became executive director of the Authority, and Ron Stewart was appointed director of engineering. The legislature authorized a bond issue to yield $8.6 million.[4]

Thus, the project definition and funding phase took four years, probably largely because of inertia in the legislative process; the cost in time devoted to the project by the planners is unknown.

[1] McCombs, "Chicago's Television College."
[2] Bretz, *Three Models for Home-Based Instructional Systems.*
[3] Maclure, "England's Open University Revolution."
[4] Belt, "Education in Kentucky—By Television."

In contrast, the Anaheim project, in which television is used for basic instruction, took less than a year to formulate.[5] The idea germinated in April 1958 when Robert Shanks and two board members saw demonstrations of the Dade County and Hagerstown projects at a National School Boards Conference. In the fall, Shanks and the Assistant Superintendent for Audiovisual Education spent several weeks touring the country to survey school TV projects. Grant applications went to The Ford Foundation and the federal government during March to April 1959.

The Anaheim case also illustrates the variations that may be encountered in the progression of steps in project implementation. The grant applications were written *after* most of the work in the planning phase had been completed and the Anaheim sponsors had already designed and costed the project.

Planning

The tasks in the planning phase break into five major areas, as grouped in Table 2. The first has to do with the objectives of the project. If objectives are to direct project design, you must refine them further. They will also direct the design of the evaluation, if there is to be one. Next you will need to plan such features of the project as producing or acquiring TV programs, providing concurrent instruction or counseling, obtaining feedback from the viewers, disseminating information about the project, and interacting with outside agencies. Then you can estimate the project's requirements for dollars and other resources. You should estimate your needs for support from the cable TV system and concurrently establish liaison with the cable TV operator, to ensure that your requirements are realistic. Finally, you should define the organization that will implement the project: Who will play what roles in planning, developing, and operating the project?

This is a large order. Nevertheless,

> success depends on having a good plan of action, from the outset, long before the electronic tubes light up.... Such planning in practice...can never be perfect or final, if only because present facts are never complete and future developments can never be certain. But choosing a destination before the trip begins and mapping the route and time-table as carefully as possible, especially through unfamiliar territory, invariably improves the journey.[6]

Like project definition and funding, planning can take varying amounts of time, depending on the degree of innovation envisaged for the project. It is not unusual for a year or more to be devoted to planning activities, although some projects have been planned in considerably less time. For example, the successful project in Hagerstown, Maryland, for providing basic instruction via television programming was planned *and* developed in about four months. But, although the planning and development phases for the Scarborough TV College in Ottawa took almost two years, this project has had rough sledding since its inception. Thus, the mere amount of time devoted to planning does not ensure initial success.

---

[5] Shanks, "The Anaheim Approach to Closed-circuit Television."
[6] Schramm et al., *The New Media.*

Development

Development tasks fall into four general areas. Most of the work will go into integrating, scheduling, and producing or acquiring the films, tapes, and other materials that the project will need. In addition, you must prepare for concurrent instruction or counseling, if any, and set up arrangements for audience feedback. A formative evaluation may run through this process to help ensure that the materials and programs are effective. You must choose channels and schedules for distributing TV programs, and users must be connected to the cable TV system. You should disseminate information about the project to potential users and others who may provide support. Tasks having to do with the organizational support of the project include obtaining and training personnel who will operate the project and setting up the intraorganizational lines of authority and responsibility.

The time required for these activities is, again, variable. The producers of the Bavarian *Telekolleg*, which offers secondary-level courses via television to adults, took a year; the *Sesame Street* staff, 20 months; and as noted above, Hagerstown required only four months to plan *and* develop its project.

Operation

The tasks to be performed in the operation phase listed on Table 2 are self-explanatory. They are largely the culmination of planning and development phases.

Total Time To Be Devoted to Implementation

Earlier we noted that the time required for project definition and funding, planning, and development phases of program implementation is rarely the sum of the times required for each of the phases. To illustrate this point, we show the implementation phases involved in producing *Sesame Street* in Fig. 2. Each of the three implementation phases is indicated by a different shading. Note that the degree of overlap among the phases allowed the entire process to be completed in a little over three years, instead of the four years that might have otherwise been required.

In contrast, the time required for implementing the Anaheim closed-circuit TV system was paced by the school year. The funding and planning phases were essentially complete in early spring. The first lesson was televised in September.

Obtaining Assistance in Implementing the Project

Special skills may be needed for funding and planning, skills that may not be available within a small school system. Even large school districts, which already have curriculum designers, cost analysts, and the like, may lack the expertise re-

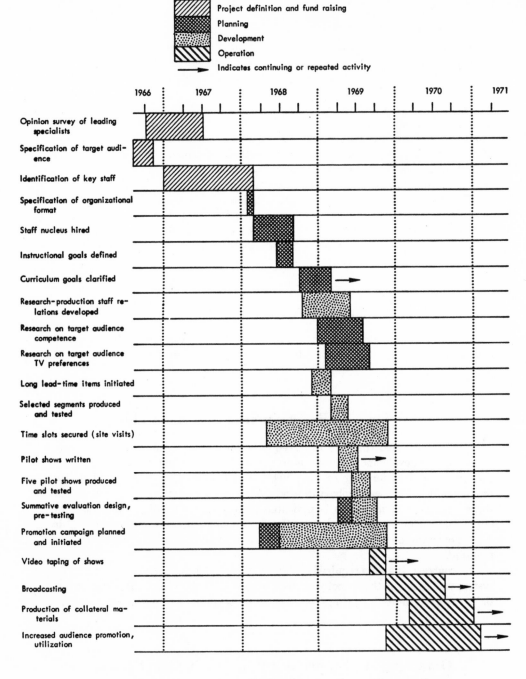

Fig. 2—Phases and tasks in the implementation of *Sesame Street* (adapted from Bretz, *Three Models for Home-Based Instructional Systems)*

quired not only to determine which uses of cable TV are best for them but also to direct the design and development of a cable TV-based education project. If this is the case, you may find it helpful to contact people in organizations that have been active in assisting schools in using TV. Such people rarely can solve your problems, but they can suggest techniques for solving them. Among such organizations are

PubliCable, National Education Association, Harold Wigren, 1201 Sixteenth Street, N.W., Washington, D.C. 20036. Telephone: (202) 833-4120. Active in assisting educators in using cable.

Joint Council on Educational Telecommunications, Frank W. Norwood, 1126 Sixteenth Street, N.W., Washington, D.C. 20036. Telephone: (202) 659-9740. Has information on cable developments in education.

Suffolk County Organization for the Promotion of Education, Roger W. Hill, Jr., Suffolk Educational Center, Stony Brook, New York 11790. Telephone: (516) 751-8500. Will advise on franchise negotiations for educational applications.

National Instructional Television Center, Bill Perrin, Box A, Bloomington, Indiana 47401. Telephone: (812) 339-2203. Has information on current uses of TV in education and on availability of recorded programming.

National Association of Educational Broadcasters, James A. Fellows or William T. Dale, 1101 Dupont Circle Building, 1346 Connecticut Avenue, N.W., Washington, D.C. 20036. Telephone: (202) 667-6000. Has long been active in use of TV and radio in education; will provide up-to-date information on current projects.

Division of Technology Development, U.S. Office of Education, Lawrence Grayson (funding) or Michael Nebbin (technical assistance), 300 Seventh Street, S.W., Washington, D.C. Telephone: (202) 755-7575 (Grayson) or (202) 755-7707 (Nebbin).

IV. PLANNING AN IN-SCHOOL PROJECT

This main chapter sets forth guidelines for the project definition and funding phase and the planning phase. The differences between planning an in-school project and planning an out-of-school project are discussed in Chap. 5. Chapter 6 treats the development and operating phases.

Project Definition and Funding

Table 2, page 200, lists the tasks to be undertaken during the planning phases in the first two columns. Even though you do not intend to request additional funding to support your project, several of the steps listed here are essential to success. In particular, the identification of the target audience and the establishment of general goals direct all subsequent activities and also supply the rationale for any request for funds. A case in point is *Sesame Street,* where the idea (applying professional TV entertainment techniques to education), the *target audience* (preschool children—ages 3 through 5), and the *key staff* (executive producers, program producers, public relations, chief education consultant, etc.) were identified before submission of the initial application for funds to finance the program.[1]

It is also important for you to determine early what your relationships with the cable TV system will be. Of course, if the franchise has already been let, you will be constrained by its terms; if not, you may be able to influence its terms in your favor, as suggested in Chap. 2.

Frequently, the people who carry out the work of project definition and funding are subsequently key figures in project design, development, and initial operation. Robert Shanks, Superintendent of Anaheim Schools, and David Snow, a member of the Anaheim Board of Education, were both instrumental in seeing the Anaheim project through to successful operation. Similarly, Len Press pushed the Kentucky project for ten years. Personal involvement and dedication by the key personnel *throughout* the implementation process are crucial to project success.

[1] Bretz, *Three Models for Home-Based Instructional Systems.*

Project definition activities should be undertaken by people who have clearly in mind an important educational problem (or problems) that they wish cable TV to help them solve and who are seriously committed to making a substantial and sustained effort to this end. They must not simply be enamoured of a new fashion or of toying with a new gadget. If the project is to undertaken at all, it should be viewed as a priority and not as a marginal activity.[2]

Some of the most successful and imaginative projects have been developed by people with backgrounds in both education and communications. Press was head of the radio-TV-films department at the University of Kentucky; Snow was an executive of an electronics firm; *Sesame Street's* Cooney had produced TV documentaries. Thus, effective use of TV in education is abetted by cross-fertilization of the two fields. This assures that the project director will be cognizant and understanding of the institutional problems he may confront and will also be aware of the potential of cable TV as an educational tool.

This leads naturally into a discussion of another task in the project definition phase—the identification of key resources—because the most important resource is a skilled and involved project director. Two other important resources are

- Other groups who may provide support or who may strengthen the project by joining it.
- Existing means of TV distribution.

Planning

We shall discuss three areas in the planning phase (refer to Table 2):[3]

- Specifying project objectives.
- Designing the project and estimating resource requirements.
- Defining an organization to conduct the project.

As will be seen, planning may require specific skills of project design and analysis. Therefore, you may want to include on the planning staff people with skills in such areas as curriculum design, design of facilities and audiovisual equipment, and resource and cost analysis. At the same time, however, it is very important to include in the planning process intermediate and ultimate users of the project. For example, if classroom teachers will be expected to introduce or follow-up the TV lessons with classroom activities, they should participate in planning tasks that bear on these activities, as well as in tasks that bear on the content of the TV lessons. Students should be included in some later tasks, such as tryout of pilot programs. And if a new institution is being established to support the project, participation of students in some of the planning tasks would also be valuable.

---

[2] Schramm et al., *The New Media.*

[3] The design of the evaluation is described in Chap. 7; most of the factors to be considered in planning the role of the cable TV system have been discussed in Chap. 2.

SPECIFYING PROJECT OBJECTIVES

Detailed objectives provide direction and coherence for all subsequent activities. In addition, they define the measures or information that will be accepted as evidence of success, and hence define the content of the summative evaluation.

The objectives of the project are actually interrelated sets of specifications of *what* is to be taught (or presented) to *whom* and *why*. The *why* of the project should have been established in the preceding phase. Now let's look at the detailed specification of the *who* and the *what*.

Identifying the Target Audience

You should know the following characteristics of the target audience:

- Ages or grade levels
- Sex
- Ethnicity
- Economic level
- Language capabilities (reading level, vocabulary, bilingual ability)
- Level of knowledge and skill in the subject areas of interest
- Numbers of students in each major category

In a single school district, these characteristics will already be known, by and large. Most schools also have routine testing programs that provide data on their students' intellectual achievement. If several school districts are involved, as in a consortium or a regional network, information on students should be collected in a central place so that channel space and program schedules may be allocated on a rational basis.

It may also be valuable to determine students' emotional needs, attitudes toward school and learning, values, and goals. Few schools have information on these aspects of their students' personalities, but such information can be very useful in directing the design of the project. It is particularly important when the project draws its target audience from a group that has the choice of using the project or not, depending on how well individual needs are met.

If television is to be used as a part of the classroom teachers' repertoire, the teachers become an influential part of the target audience. They should, therefore, view the project as supporting their efforts; to this end, their help should be elicited during the planning and development phases. The involvement of teachers in the planning stage may partly explain the success of the Hagerstown venture.

> In the summer of 1956 nearly 100 teachers, principals, instructional supervisors, Parent-Teacher Association (PTA) leaders, and consultants representing various subject-matter areas, geographical regions, and community groups met in a six-week workshop to plan the televised curriculum.[4]

Although teachers are involved in the specification of materials and approaches to be used, you must be careful that their needs and biases do not overshadow the

[4] S. Wade, "Hagerstown."

needs of the ultimate audience, the students themselves. The *Sesame Street* experience, which verified that preschoolers have attention spans well in excess of the 15 minutes often cited by educators, is instructive in this regard.

Choosing Program Content [5]

First to consider in the choice of content for the televised portions of the instruction is whether television is an appropriate medium for the subject matter. Almost any subject matter can be taught at least in part by television; the question is whether that part is significant enough to warrant the medium. Subjects largely inappropriate for television will fall mainly in vocational areas (building trades, repair of equipment or machinery, service occupations) or in areas requiring large amounts of student performance, especially if interpersonal or team skills are required.

To do the television medium justice, however, we should not consider its appropriateness only in terms of activities for which it can *substitute* for classroom instruction. A wide variety of audiovisual experiences can be provided the student *only* through the media of television or film, and many others would be so costly to supply by direct experience that they are inaccessible for all practical purposes.

Thus, film and television programs are not only appropriate for a large part of conventional classroom instruction, but they can also greatly expand the range of instruction that can be given in the classroom. In fact, the potentialities of film and television for instruction have not begun to be realized. The problem in choice of programming for television is not one of finding subjects for which television is appropriate, then, but of narrowing the selection to those subjects for which the use of the medium is most practicable. What is practicable will depend on the fundamental goals of your project, especially on whether programming exists (or can be produced within available resources) that meets the needs and interests of the target audience.

Other considerations that hinge on the general goals of the project will affect the choice of subject matter. These goals, illustrated in the companion volume, are

- To make education physically more accessible to students.
- To provide additional services that cannot readily be provided by other means.
- To improve the quality of education.
- To decrease the unit cost of education.

To increase the physical accessibility of education, subject matter must appeal to a large enough segment of the target audience to ensure favorable per-student returns for investments in programming. To provide additional services, subject matter will differ from that routinely presented in the classroom. For example, foreign languages are often taught by television because of the shortage and cost of proficient teachers. The quality of instruction can be improved by enrichment or other supplementary uses or by freeing the classroom teacher to give students individual attention. Unit costs can be decreased if television programs are used as

[5] Part III of this book discusses the uses of television in education at greater length.

a substitute for some teacher activities. For example, in some courses with large enrollments the University of California at Berkeley televises live lectures to students in adjoining classrooms. Other considerations of practicability are discussed at greater length below.

DESIGNING THE PROJECT

Goal definition sets the stage for project development; design lays out the best way for the cable TV system to support these goals. The design process can be back-of-the-envelope calculations and flow diagrams or complex computer-driven models. But simple or complex, it will involve several steps, interrelated as shown on Fig. 3. The figure is divided horizontally into three parts. At the top are inputs from subject matter, the student population, the school system, and sources of recorded TV programming. At the bottom are outputs such as requirements for resources of various kinds. The design process, in the center, relates inputs to outputs by means of the steps shown.

Steps 1 and 2

We have already described Step 1 (analyze target audience) and Step 2 (analyze subject matter) as parts of the detailed specification of objectives.

Step 3: Specify School Requirements

Items that should be considered here direct several subsequent decisions. Some items are

- How will the schedule of television programs be fitted into the school calendar and school day? Conversely, to what extent can the school day be adjusted to accommodate the television schedule?
- Do existing education codes or union rules limit the roles that television can play in the school setting? Can or should these limitations be changed, and, if so, how?
- Are there institutional limitations on the use of regular faculty on the project? What incentives will encourage the faculty to participate?
- Will the project compete with other established agencies within the schools such as the libraries? Can such conflicts be ameliorated or resolved?

Scheduling will be a problem if your project must fit into a daily operation that is rigidly scheduled to meet other requirements. For example, in junior and senior high schools, departmentalization of subject matter makes scheduling difficult. In the usual arrangement, ninth-grade English, for example, is taught during every period in the school day. Under these conditions, Bretz has estimated that 13 channels would be required to provide 16 percent of instructional time for a secondary school system.[6] At the elementary level this problem does not arise because the

---

[6] "Potential Uses of Cable."

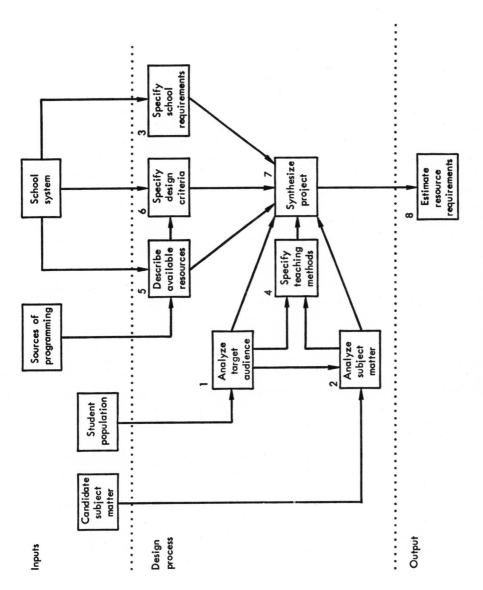

Inputs

Design
process

Output

Fig. 3—Steps in the project design process

classroom teacher can arrange his day as needed to accommodate the televised programs.

Several states have mandated upper limits on student-teacher ratios at various grade levels. These limits may restrict the extent to which you can substitute televised lessons for classroom teachers. Sometimes you can use teacher aides to decrease the student-teacher ratio without a proportionate increase in cost; however, the teacher union in Addison, Illinois, has charged that it is illegal to use only a paraprofessional in a classroom receiving a televised lesson (as of this writing, the case had not been resolved.)[7] Sometimes projects can be declared to be experimental and therefore not subject to regulations imposed on the normal classroom.

You should also determine the roles of the regular faculty in the project in light of institutional requirements as well as of teacher interests. Some teacher unions, in their battle against merit pay, will not allow teachers performing special functions (such as producing televised programs) to be paid more than other teachers at the same rung of the salary schedule. This makes it hard to figure out what incentives will encourage faculty participation.

If regular teachers record their lessons on videotape, they may feel they are doing away with their own jobs. For them, additional compensation seems justified. Some school systems that make extensive use of locally produced programming prefer to make minimal use of videotape, with a consequent increase in the cost of programming. For example, the University of Southern California has requested that each user of the lectures to be televised over its ITFS network erase at the end of each semester all videotapes that have been made for the convenience of the students.

Librarians or directors of media services quite naturally view cable TV as directly competitive. You should try to make the services provided by such agencies an integral part of the project during the planning and development phases. If personnel in these agencies have active roles, they will be more likely to contribute their support and expertise.

Step 4: Specify Teaching Methods

Returning to Fig. 2, the next step is to specify the teaching methods most appropriate to the target audience and the subject matter. By far the commonest methods used in instructional television are those in which the viewer is ostensibly passive; that is, he is not asked to respond overtly during the program. Even with this restriction, wide variations in format are possible. The stand-up lecture, lecture with demonstration, documentary film, instructional film, and even the magazine format of *Laugh In* have all been used as teaching tools. These methods can be highly effective (witness *Sesame Street* and *Telekolleg),* especially if you are careful to match the content and format of the presentations to the needs and wishes of the target audience.

The companion volume to this one discusses a number of other possibilities, including using the two-way feature of cable TV.

[7] "Teacherless Classes," *Nation's Schools.*

Step 5: Describe Available Resources

Resources that are already available to support the project will strongly influence its design, as will the resources you believe you can obtain. Aside from dollars, essential resources are

- Personnel
- Television programming (software)
- TV system hardware
- Student time available; student locations
- Materials for concurrent instruction

We have already talked about the essential characteristics of key personnel. In addition, librarians, audiovisual specialists, and curriculum specialists will be valuable in selecting programming or in local program production. Of course, if your school system is already using television in instruction, engineers, technicians, and studio crew will also be available, as may be teachers who are skilled in producing instructional programming.

During the planning phase, you should contact the large repositories of educational and instructional audiovisual materials to find out whether you can meet a large part of your programming needs with existing software. A list of these repositories is given in App. B. Unfortunately, there are no rating services, and only a few of the repositories supply even search services. This means that to find programming you may need to use consultants.

A checklist is helpful in this initial search: It should contain

- The subject areas of interest.
- The primary function of the television programming; i.e., is it supplementary or basic to instruction?
- The maturity levels of the target audience.
- The ethnicity of the target audience.
- The state of prior knowledge or performance capabilities of the target audience in the subject areas.

If you are lucky, good quality programming that meets your needs will be available. More likely, appropriate programming will not be available, especially if it is for basic instruction. In this event, you will need to plan on either producing your own programming, sharing production costs in a consortium of schools, or using some mix of locally produced and off-the-shelf programming.

Some school systems already have television receivers, videotape recorders, interior school wiring, receiving antennas, and production studios. The Willingboro School District, for example, already had an extensive CCTV system, including its own facilities and equipment for program production, when it decided to use the newly franchised cable system to involve parents through school-to-home programming.[8]

Naturally, you will take into consideration any facilities or equipment of this kind during the planning stages. In rare instances, existing resources will be suffi-

---

[8] Reuben, "Using Cable Television To Involve Parents."

cient to support the project; more likely, you will need to add facilities and equipment. Bear in mind that the capital costs of the project for facilities and equipment will undoubtedly be dwarfed in the long run by the recurring costs of operation, particularly if you are thinking about an appreciable amount of local program origination. For example, after the Hagerstown project had been running for eight years, the capital costs were less than 20 percent of the total recurring costs to that date. A number of school systems have been caught short because only the capital costs were considered during the planning phases.

Because the cable TV system will reach many places besides the schools, you should consider what role receivers outside the schools will play in the project. Receivers for program distribution may be in the home or in new facilities built for receiving the programs.

In deciding what proportion of the effort will be home-directed and what classroom-directed, be realistic about the school's ability to attract a home audience of sufficient size to justify the resources needed for appropriate programming. Consider also whether your target audience already has cable service at home. (It is likely that poorer families will not.) On the other hand, if you plan to use the two-way feature, it will be so expensive that you will probably want to equip only selected home receivers, or receivers in special facilities with return terminals.

Step 6: Specify Design Criteria

Just before synthesizing the project you should specify criteria to guide the process (Fig. 2, p. 204). Examples of such criteria are

- Lowest cost over a given period, say, ten years.
- Lowest capital cost.
- Lowest recurring cost.
- Maximum utilization of television in instruction.
- Lowest cost of instruction per student in project subjects over a given period.
- Maximum utilization of available personnel.
- Maximum utilization of advanced technology.
- Maximum likelihood of additional funding.

The set of criteria you choose and the rankings you give them should reflect the financial and political realities of your situation. For example, if building funds are short you might use existing facilities to the greatest extent possible, even though over the long run this may not lead to the lowest cost project. Or if the school budget is in trouble because of the level of teacher salaries, lowering the cost of instruction per student may override maximum utilization of existing personnel.

Step 7: Synthesize Project

You are now ready to synthesize the project. Many variants of project characteristics can be contemplated, and many will interact with one another in a variety of ways. For example, as shown previously, the number of channels required is not directly determined by the number of lessons to be transmitted daily.

Because of the large number of ways in which projects may be synthesized, we cannot outline a single best procedure. What is best is what is best in your situation, which cannot be defined a priori. Here we can only provide two very different examples of projects that are currently in operation in existing school systems—Hagerstown, Maryland, and Norwood, Ohio. These examples will give you a feeling for how different modes of operation are supported in different school systems.

Table 3 shows features of the Hagerstown and Norwood projects. Both are closed-circuit systems that lease six channels from the telephone company. Two features so differentiate the projects that the recurring costs differ by more than a factor of six:

- The primary use of TV in Hagerstown is instructional; in Norwood, teachers use programs at their discretion to enrich regular classroom teaching.
- Hagerstown produces nearly all of its programming. Norwood fills one channel with the programming offered by WCET, the educational TV station, and one with programming provided by the Southwestern Ohio Instructional Television Association; classroom teachers use the other four channels to access films and tapes stored in the Norwood media center.

Table 3

COMPARISON OF THE HAGERSTOWN AND NORWOOD PROGRAMS

	Hagerstown	Norwood
Schools served	45	11
Primary use of TV	Instruction	Enrichment
Classrooms with sets	437	100
Channels used	6	6
Miles of cable leased	125	5
Grades served	1 to 12	K to 12
Lessons transmitted per week	150	381
Hours transmitted per week	75	108
Lessons produced locally per week	141	Few
Personnel	87	5
Capital cost	$448,000	$75,000[a]
Recurring cost	$324,864	$55,000[a]

[a]Rough estimate.

Thus, although Norwood transmits more hours of programming weekly (and more than twice as many individual lessons) than Hagerstown, the project makes much lighter demands on school system resources because of its objectives and because of the way in which it is conducted. This illustrates how heavily project costs and requirements for other resources depend on the details of project specification. This is why relatively complete project design is desirable in the planning stages.

The results of project synthesis will be

- A schedule of televised program transmissions.
- The number of channels required.
- The number and locations of TV sets required.
- A schedule of local program production, if any.
- A list of facilities and equipment for local program production, if any.
- A strategy for the use of televised programs.
- A strategy for publicizing the project and obtaining outside support.
- Tables of personnel needs.

Schedule of Program Transmissions and Number of Channels. The number of subjects to be taught and the degree to which the school schedule will accommodate the TV schedule are all important here. Although under some conditions more than a dozen channels would be needed to present basic instruction in secondary school, if the classroom schedule can be rearranged to accommodate the televised lessons, six channels is more than sufficient. This is illustrated by Hagerstown's planned use of channels for 1972-1973, shown in Fig. 4. Hagerstown achieves efficiency of channel use by unusual classroom scheduling, as illustrated by the 8th-grade schedule for 1972-1973, shown in Table 4. Note that all eighth graders throughout the entire Washington County school system take TV-science, social studies, math, English, and art at the same time.

Of course, if you plan to use the cable TV system to transmit audiovisual materials immediately upon a teacher's request, no prior scheduling can be contemplated. In this event, the number of channels required will depend on the number of teachers who will be using the system simultaneously and the number of titles in the media center. A scheme of this sort has been tried in Ottawa; on the basis of their experience, one channel would serve about 21 classrooms from a library of 2800 titles.[9]

Number and Location of TV Sets. The number of sets required in each room depends on the room size and the size of the picture tube. Bretz gives the following rule of thumb: The maximum viewing distance from the television set is 12 times the width of the screen, and the optimum distance is 6 times the width of the screen.[10] Thus, a 21-inch set should not be viewed from farther than 21 feet, and preferably from about 10 feet.

You may not be planning to use television in some subject areas above the elementary level and therefore will install television sets only in rooms where those subjects are taught. You can realize economies here if you can reschedule the use of rooms to minimize the number needing sets.

Planning Local Program Production. The companion volume presents strong arguments against local production of TV programming. You may decide, however, that you have sufficient resources to create programs of high quality, that you can transmit programs with strong local appeal, or that local production is the only way to obtain programs that will fit your needs. Programs of poor *production* quality can still be effective as teaching devices if they make up in imagination and interest for their lack of slickness. Another way to ensure that programs are used is to make them an integral part of the reward structure of the school systems, for

[9] McLaughlin et al., *Educational Television on Demand.*
[10] *Will My Visual Be Visible?*

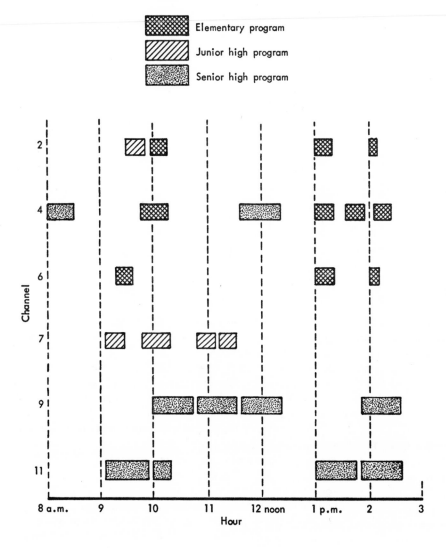

Fig. 4—One day of channel utilization, Hagerstown, 1972-1973

Table 4

PROPOSED WASHINGTON COUNTY TV SCHEDULE, 1972–1973

(8th Grade)

Hour	Day				
	Monday	Tuesday	Wednesday	Thursday	Friday
9:05–9:27	Science 8 Ch. 7	U.S. Studies 1 Ch. 7	Science 8 Ch. 7	U.S. Studies 1 Ch. 7	Science 8 Ch. 7
9:28–9:50	Math 8B Ch. 7	English 8 Ch. 7	Math 8B Ch. 7	English 8 Ch. 7	English 8 Ch. 7
	Art 8 Ch. 7		News Lab Ch. 2		
1:50–2:12		Art 8 Ch. 7			

example, to make a student's grades a least partly contingent on his mastery of program content or to award credits for TV courses. Such an arrangement should not be an excuse for tedious programming.[11]

No matter how much programming is to be produced locally, almost certainly you will need to rent or purchase some. Two percent of the programming used in Hagerstown, for example, came from outside sources in 1965. You will need people who are familiar with sources and who can judge quality in selecting useful programming. If you do not already have such people, you may need to train someone for the task. You must also have a projection room, suitably equipped with film chains and videotape recorders. This room might be in the studios of the cable TV operator or in a school-operated projection facility.

Strategy for Program Utilization. The most vital aspect of utilization is continuous in-service training for teachers and aides. They must learn not only how to use the television sets and to follow the program schedules but, more important, how to provide the preparatory and follow-up activities that will reinforce the programs. You should make arrangements so that classroom teachers can be trained and can plan and carry out support activities within the regular teaching day.

You may want to go beyond traditional utilization activities by training teachers to use the TV lessons as springboards for individual instructional activities, group projects, or guided classroom discussions. In fact, provision of basic curriculum via TV may give teachers and aides additional time to develop their own ideas of what materials and methods work best for their students. To make effective use of such time, they must learn what materials and methods are available and how to use them.

In addition, regular in-service training will

- Serve to acquaint new teachers with the project.

[11] Guidelines for planning local production capabilities are discussed in several recent works. See Quick and Wolf, *Small-Studio Video Tape Production.*

- Provide opportunities for implementing new approaches for program use as they become apparent.
- Provide teachers with opportunities to make suggestions for improving the project.

The more extensive your utilization program is, the more training you will need to provide people who will prepare the materials relating the programs to the basic curriculum. In addition to the usual lesson guides and study guides, you may want to use programmed texts to reinforce the cognitive content of the TV lesson, audio tapes to provide additional drill in, say, word recognition or pronunciation, or book lists for related reading.

Teachers will need lesson guides to make best use of the programs. Such lesson guides should

- State the learning objectives for each program.
- Relate the lesson to the curriculum sequence and other instructional activities.
- Provide directions on conducting preparatory or follow-up activities.
- Provide means (such as an occasional short quiz or an end-of-series test) for finding out whether students have attained the learning objectives.

Before the project gets under way, you must allow for space and equipment for producing such materials and should procure initial stocks of supplies.

Program utilization can be further enhanced if students can respond directly to the program during its presentation. In most cases, responses will be entered on workbooks or answer sheets. You may, however, have chosen to make immediate use of student responses during the program by sending responses back to the projection room or studio. The number of response devices and response monitors required will depend on the number of students using the devices at a given time. Once the response system has been installed, personnel should be assigned to maintain it and to replace equipment as needed.

Project monitoring[12] mechanisms are desirable for several purposes:

- Spotting deficiencies in time to remedy them before the project is severely damaged.
- Providing a vehicle for expression of audience reaction.
- Providing a vehicle for suggestions for project improvement.
- Providing data for continuing project evaluation.

Some monitoring mechanisms are inexpensive. For example, unannounced classroom observation followed by interview of the classroom teacher can be used to spot-check usage of particular programs. Short, anonymous questionnaires can be administered to teachers and students involved in the project. And teachers can occasionally be given release time to discuss with project managers the content and format of the televised presentations, the operation of cable TV system, the scheduling of presentations, and the production of materials to enhance program utilization.

---

[12] This is sometimes called process evaluation. We prefer to consider it an integral part of an ongoing project, however.

You will also want a continuous check on

- The effectiveness of classroom teachers in implementing the project.
- The internal workings of the supporting organization.
- The television system itself: sets, picture quality, and cable reliability.

In his case studies, Wagner found that

> Few school systems had adequate back-up equipment of any type, most schools had an inadequate number of outlets and receivers, preventive maintenance procedures rarely were in effect, and emergency maintenance services were unable to respond quickly enough to prevent "lost" instruction in the classroom.[13]

You should include in the monitoring process a program for gathering test data or other measures to determine the effectiveness of the project. You will want to consider what kinds of measures will be gathered and how. Standardized achievement tests are available in most subject areas, but they are inadequate as measures of effectiveness of specific short-run instructional projects. You may prefer to devise and administer tests composed of items keyed to the instructional objectives of the project. These can be (but need not be) criterion-referenced measures; at least they should be scrutinized by your testing group for reliability and face validity. You may want to administer the tests via television to eliminate, as far as possible, the effects of individual teacher biases. Further discussion of appropriate measures will be found in Chap. 7.

Plan Publicity and Liaison Activities. In an in-school project there will be minimal need for publicity, if teachers have been adequately prepared for classroom utilization. If not, no amount of publicity will compensate for this deficiency.

Agencies independent of your school system will contribute to the project. For example, the cable system operator will connect the schools to the system and provide technical consultation. If your schools belong to a consortium, other members will play important roles in project implementation. And, if outside funding will be used, sponsors will want to monitor the progress of the project.

Maintaining effective communications among independent agencies is difficult and is thus often neglected. It is easier to pay attention to problems within your own organization than to try to understand and accommodate the concerns of outsiders whose goals are likely to conflict with yours. Therefore, you should anticipate difficulties and make arrangements for resolving them from the outset. In planning, two steps are important:

- Determine the agencies with whom you expect you'll need or want close relations.
- Select people within your organization whose job will be to maintain harmonious relations with these agencies.

The less closely your project is tied to an existing school system, the more it may depend on the support of outside agencies, and the more importance the liaison function will assume.

---

[13] *A Study of Systemic Resistances to Utilization of ITV in Public School Systems.*

Tables of Personnel Needs. You can attain considerable flexibility in the use of personnel by assigning them to various functions part time. See the example given in Table 5, which compares the personnel used for local program production in Hagerstown and Santa Ana; the number of programs produced in Santa Ana was less than a seventh of that in Hagerstown.[14]

Table 5

PERSONNEL ASSIGNED TO LOCAL PROGRAM PRODUCTION

	Hagerstown	Santa Ana
Lessons per week produced	141	20
Function	*Personnel Assigned*	
Production manager	2	0
Producer-directors	8	2
Assistant directors	1	0
Supervisor of television content ..	1	Done by project director
Telelesson teachers	34	5
Studio crew	16	1.5
Graphic artists	4	1
Sets and props	1	Done by graphic artists
Lighting	Done by directors	Done by directors
Librarians	Done by telelesson teachers	Done by telelesson teachers
Film	Done by technicians	Done by telelesson teachers
Photographic technician	1	0
Chief engineer	1	1
Assistant chief engineer	1	0
Technicians	8	2
Guest relations	1	Done by project director
Total	79	12.5

Step 8: Analyze Project Costs

The last step in the design cycle is an analysis of the costs of the project (Fig. 2). For assistance in planning, you should provide a moderate amount of detail on project costs. If you have designed your project as suggested, you will be in a good position to derive the needed details.

The first step in cost analysis is to decide how costs will be categorized. It is generally convenient for planning to use two major cost categories: acquisition costs (the one-time costs of acquiring the project) and operating costs (the recurring costs for running the project once it has been acquired). Within these major categories,

[14] Bretz, "Closed-Circuit ITV Logistics."

resources are grouped into subcategories—personnel, facilities, equipment, and materials—because these usually relate to categories in the regular school budget.

We have chosen to further categorize resources by project function, as shown on Tables 6 and 7, because the functional grouping facilitates planning and evaluation. Beneath each function are listed the categories of resources that will be used for that function. Under the column headed "Item" are brief descriptions of the resources falling into each category.

The columns headed "Cost-generating Basis" display the units on which the total cost of a particular task depends. For example, the total cost of cable drops to schools depends on the cost per school drop and the number of schools in the project. Hence, the individual schools in the project are the basis for generating these costs. If a resource requirement is generated merely by the existence of the project (as in the case of function 1), the project itself is the cost-generating basis, and a check appears under the column headed "Project." If a resource requirement is generated by each lesson transmitted, a check appears under the column headed "Lesson." Other cost-generating bases are listed under the column headed "Miscellaneous." Note that in a number of instances a category has more than one cost-generating basis. This occurs in cases such as local program production when a minimum resource (such as a producer-director) is required independently of the level of the activity in question, but more resources will be needed above the minimum in proportion to the level of the activity.

The first two functions of both tables are mandatory for any cable TV education project. Someone must run the project and the project must involve a tie-in to a cable TV system. Each school that participates in the project will require a drop (sometimes installed free by the cable system operator) from the main cable and may make use of dedicated lines installed only for school use along with the cable. Within the participating schools, TV receivers in classrooms, auditoriums, and offices must be wired to the receiving installation.

The project can become very costly if much local programming is produced and if high-quality production is a goal. The resources required for local programming will be generated from two bases. The existence of the requirement for local programming will, of itself, generate minimal requirements for facilities (studio and materials storage), personnel (primarily a producer-director and a cameraman), and equipment (primarily cameras, videotape recorders, and TV monitors). Requirements beyond the minimum depend on the number of lessons produced per week and on the complexity of production contemplated. The items listed for operating cost categories under this function are self-explanatory, in light of the above discussion.

The last two functions, publicity and evaluation, rely primarily on personnel. Their cost is almost entirely determined by the scope of the effort planned specifically for these tasks, not on characteristics of the project itself. The exception is the cost of tests and other instruments for measuring the impact of the project on student learning or attitude change. During project acquisition, planning and monitoring activities will take place in both functional areas. During project operation, lines of communication will be maintained and the summative evaluation will be conducted.

Table 6

ACQUISITION COST CATEGORIES FOR AN IN-SCHOOL CABLE TV PROJECT[a]

Function and Cost Category	Item[b]	Cost-generating Basis		
		Project	Lesson	Miscellaneous
1. Project planning and development Personnel	Plan and develop project			
2. Acquire cable transmission system Equipment	School drops Dedicated lines Intraschool wiring TV receivers			School School School Classroom
3. Prepare for rented or purchased programming Personnel Facilities Equipment Materials	Provide preservice training for ordering and screening materials Projection room Materials storage Projection equipment Purchase of initial programming	✓ ✓ ✓ ✓	✓ ✓ ✓ ✓ ✓	
4. Develop initial local programming Personnel Facilities Equipment Materials	Provide preservice training for production personnel Write and produce initial programming Production studio Materials storage Production equipment Production materials	✓ ✓ ✓ ✓ ✓	✓ ✓ ✓ ✓ ✓ ✓	
5. Prepare for classroom utilization Personnel Facilities Equipment Materials	Provide preservice training for teachers and aides who will utilize the program Provide preservice training for preparing utilization materials Write and produce initial utilization materials Shop for producing related materials TV monitors Student response devices Student response monitors Production of utilization materials Stock for student materials Stock for teacher materials	✓ ✓	✓ ✓ ✓ ✓	Teacher Classroom Student Student Student Teacher
6. Initial publicity Personnel	Establish liaison with independent, involved groups Plan public relations strategy	✓ ✓		
7. Formative evaluation and research Personnel Materials	Plan evaluation; gather baseline data Test and measurement materials	✓		Student
8. Other services				

[a]Adapted from Haggart, *Program Cost Analysis*.

[b]For personnel costs, salaries (including fringe benefits) should be prorated to the time required to accomplish the tasks listed.

Table 7

OPERATING COST CATEGORIES FOR AN IN-SCHOOL CABLE TV PROJECT[a]

Function and Cost Category	Item	Cost-generating Basis		
		Project	Lesson	Miscellaneous
1. Project administration				
Personnel	Direct and plan project	✓		
2. Operation, maintenance, and replacement of the cable transmission system				
Personnel	Maintain in-school wiring			School
	Maintain TV receivers			Classroom
Equipment	Channel rental			School
	TV receiver replacement			Classroom
3. Rent or purchase programming				
Personnel	Order and screen programs	✓	✓	
	Schedule and project programs	✓	✓	
	Make duplicates	✓	✓	
	Maintain projection facilities	✓		
	Maintain projection equipment		✓	
Equipment	Projection equipment replacement		✓	
Materials	Rental or purchase of materials		✓	
	Stock for duplication		✓	
4. Produce local programming				
Personnel	Prepare and produce local programming	✓	✓	
	Provide in-service training for production personnel	✓	✓	
	Maintain production studio	✓		
	Maintain production equipment		✓	
Equipment	Production equipment replacement		✓	
Materials	Stock for production		✓	
5. Classroom utilization				
Personnel	Prepare and produce utilization materials		✓	
	Provide in-service training for teachers and aides who will utilize program			Teacher
	Maintain TV monitors			Classroom
	Maintain student response devices			Student
	Maintain monitors for student response			Classroom
	Maintain shop for producing utilization materials	✓		
Equipment	TV monitor replacement			Classroom
	Student response-device replacement			Student
	Replacement of monitors for student response			Classroom
Materials	Stocks for producing utilization materials		✓	
6. Publicity				
Personnel	Maintain liaison with independent, involved groups	✓		
	Maintain public relations	✓		
7. Evaluation				
Personnel	Monitor project progress	✓		
	Conduct summative evaluation	✓		
Materials	Test and measurement materials			Student
8. Other services				

[a]Adapted from Haggart, *Program Cost Analysis.*

DEFINING ORGANIZATIONAL STRUCTURE

Returning to Table 2, p. 200, the last planning task is to define the structure of an organization for conducting the project. This structure will begin to emerge during the planning phase. You should consider what it should be, rather explicitly, to ensure that slots are filled in time. This will protect the health of the project and the project manager (perhaps yourself), who might otherwise try to perform so many of the functions that either he or the project collapses. Of course, if the project is small, only one or two people may be needed.

There are several bases on which an organizational structure can be formed. Whatever is used will tend to direct the attention of project personnel to those areas the structure delineates as functionally important. For example, if people responsible for classroom use are grouped under an organizational entity titled "Program Production," their classroom concerns may be subordinated to the concerns of production. Thus, your organizational structure should reflect the priorities you attach to the various functions of the project.

In addition to components associated with the operation of the project, the following support components are needed:

- A management component responsible for setting overall goals, securing funds and other resources, and controlling the operation of the project.
- A development component responsible for planning the project, conducting research, and preparing or locating initial materials and equipment.
- A budgeting component responsible for analyzing costs, projecting expenditures, and allocating funds and personnel.
- An evaluation component responsible for devising measures to assess the effects of the project, for collecting pertinent data, and for reporting results.

You should follow several general principles in addition to reflecting the priorities of various functions. First, the definition of roles and responsibilities should be flexible enough to make efficient use of talents; organization charts should be redrawn frequently. Second, each component should be large enough for the task confronting it, but not so large that communication within it is difficult. Third, the project team should include people who are problem-oriented, technology-oriented, user-oriented, and systems-oriented. Each of these orientations has its role to play; some may be combined in a single person.[15]

Many of these people may be drawn from your present school staff, especially in the areas of

- Project administration
- Ordering off-the-shelf software
- Classroom use
- Evaluation

Remember, however, that such personnel will contribute more effectively to the project if they have preservice training.

---

[15] For a more complete discussion of organizational structure, see Klaus, *Instructional Innovation and Individualization.*

If the project departs significantly from current school practice, you may need to hire consultants to help you get started or specialists to meet your continuing needs. You are most likely to have to hire new personnel for such functions as production and projection of programs and maintaining the portions of the cable TV system that belong to the schools. Consultants can help you map your dissemination strategy, design the project, and set up your program and materials production facilities.

V. PLANNING AN OUT-OF-SCHOOL PROJECT

The steps for planning an out-of-school cable TV project are the same as those for planning an in-school project, but the areas of emphasis are different. Because the target audience is not conveniently available in an out-of-school project, you may need to put much more effort into establishing effective links with it.

Project definition is particularly important for projects that are not supported by existing educational institutions. In such instances, almost without exception, you must convince the potential sponsor that the project is worthy of support. Therefore, you must identify the target audience and its needs and satisfy yourself and your sponsor that the project will appeal to the audience and be successful in meeting its needs. Surveys of the audience will be useful at this stage.

For example, *Sesame Street* grew out of a study, supported by the Carnegie Corporation, of the use of TV for the education of children of preschool age. Carnegie was concerned about the educational deficiencies of children from poor environments and the apparent failure of programs of compensatory education. They were also aware of the general agreement among psychologists and educators that the early years are critical in intellectual development, and that by the time a child reaches the first grade his potential for academic growth has largely been set. Thus, using TV to stimulate the intellectual development of preschoolers seemed a convenient and practical way to correct the problems of the educationally disadvantaged.

Carnegie commissioned Joan Ganz Cooney to study the problem. Cooney spent ten months surveying opinions of leading specialists and writing the initial proposal for a TV-based program for preschoolers.[1] Although the proposal was not submitted until February 1968, it was preceded by a study by Cooney in 1966;[2] thus, the idea had been developing for several years. The proposal led to funding broken down as follows:[3]

---

[1] *Television for Preschool Children: A Proposal*, February 19, 1968.
[2] *Potential Uses of Television.*
[3] Richard DeFelice, internal memorandum of Children's Television Workshop, October 18, 1971.

Source	Funding ($ thousands)
Department of Health, Education, and Welfare	4000
Carnegie Corporation	1500
Corporation for Public Broadcasting	626
Ford Foundation	1538
Learning Resources Institute	150
John and Mary R. Markle Foundation	250
Miscellaneous	58
Total	8122

In contrast, Chicago's TV College sprang into existence almost overnight. Looking back, there does not appear to have been any project definition or planning phase, despite the fact that The Ford Foundation added a grant of $475,000 to the Chicago Board of Education's $600,000 to finance the first three years of the project. Planning began on July 1, 1956; broadcasting of courses was under way when the fall semester started.[4] Thus, only two months went into the planning *and* development phases of the project. Most of the work concentrated on preparation for program production. TV College required little planning because it almost exactly parallels the course offerings of Chicago's well-established community college system. The largest departure from classroom practice was the careful preparation of the television programs and of the accompanying study guides. Although the sponsors did not know whether there was an audience with enough motivation to take the televised courses, the provision of college credit for these courses helped ensure that enough students would be attracted to make the venture economically attractive.

After the funding phase, you may need to devote considerable effort to defining the target audience specifically, determining its precise needs, and discovering what will appeal in the way of TV programming.

The potential audience for *Sesame Street* is around 12 million children between the ages of 3 and 5. Nearly all of these children have television sets in their homes and watch television for over a third of their total waking hours, on the average.[5] This potentially huge audience meant the Children's Television Workshop (CTW) could afford to put relatively large amounts of money into planning and development, which they did (about $1.7 million). Over $600,000 went into preproduction research and postproduction studies, many of which were directed at determining the viewing propensities of the target audience.[6]

For projects directed at a noncaptive audience, research of this kind can be invaluable in the long run. Lack of such research can mean that the project does not reach the audience for which it was intended. For example, the Open University is supposed to appeal to blue-collar workers but to have

all the social trappings of conventional university education, and all the social benefits which it evolved on degree holders.... It had been intended to weight the recruitment in favor of those who had left school early and were in manual or semi-skilled occupations.... In practice the first 25,000

---

[4] Erickson and Chausow, *Chicago's TV College.*
[5] Rubinstein et al., *Television and Social Behavior.*
[6] Connell, "How Sesame Street Raised and Spent Eight Million Dollars."

applications only brought in 601 workers in manufacturing industry and 769 shopkeepers and workers in sales and service industries.[7]

A similar difficulty arose with Germany's *Telekolleg,* which failed to attract women in the numbers desired. We can only guess at the reasons for these discrepancies between goals and realizations. Looking at the course offerings of both *Tellekolleg* and the Open University, however, one might conjecture that they may seem too academic to a population that has already been turned off by traditional education.

Telekolleg was preceded by a number of general surveys of the potential need for its services, but apparently no effort was made to determine what the intended audience would actually watch. Such a determination was the strong suit of *Sesame Street.*

Utilization functions are especially important in out-of-school projects because they provide the only routes by which the project can be adapted to the needs and desires of the target audience. Without them, you may very well be transmitting to dark sets.

Utilization can take four forms, each of which is valuable:

- Concurrent instruction, counseling, or other activities that reinforce the programming and enhance motivation.
- Monitoring the audience to determine the effectiveness of the programming.
- Ascertaining the adequacy of TV distribution and reception facilities.
- Publicizing the project to the intended audience.

The last may well be indispensable. The 1971-1972 *Sesame Street* budget for public information was, for example, nearly $700,000, or 6 percent of the total expenditures for that year. TV College advertises through mailed brochures, previews of course offerings broadcast over Chicago's educational television station, and press releases. The cost of these activities is not called out separately in the budget.

For *Sesame Street,* the monitoring and feedback function is carried out by the

CTW field-service staff in each of 14 major cities where broadbased community advisory councils and viewing centers had been created. This staff encourages the use of the *Sesame Street* magazine and similar materials by individuals and organizations, such as teachers and community volunteers, day-care nursery programs, kindergartens, parent-education programs, teenage-vocational-training programs, special programs for the retarded or handicapped, and teacher and paraprofessional training programs. Feedback to the field-services staff is channeled back to the CTW research and production staff, and thus contributes to continuous modification and refinement of show material.[8]

Less ambitious schemes are also useful. *Telekolleg* works closely with several institutes that devote substantial effort to research and evaluation; their work provides the program's planners and producers with excellent feedback in such areas as analysis of student population and effectiveness of instructional method.[9]

[7] Maclure, "England's Open University Revolution."
[8] Bretz, *Three Models for Home-Based Instructional Systems.*
[9] Ibid.

At Chicago TV College, the teacher schedules telephone hours immediately after telecasts to answer questions from credit students.

A number of successful out-of-school projects have provided concurrent instruction by means other than TV programming. This has several functions:

- It reinforces the televised lessons.
- It provides another means for communication between project staff and the target audience.
- It increases the involvement of the target audience.

Teachers in projects that offer courses for credit usually grade and comment on homework and exams by mail and make some provisions for conferences with the students. In addition, TV College has lab sessions at a branch campus every few weeks for some science courses; *Telekolleg* sends out phonograph records to accompany English courses; and the Open University sends out radio programs, lab kits, and impressive printed materials to accompany its TV offerings; it also requires students to participate in three-week on-campus sessions each summer.

Out-of-school projects using broadcast television are directed primarily to TV receivers in the home. In most U.S. communities, however, the cable system will not reach the majority of homes, and those left out may be the ones you particularly want to reach, those in low-income neighborhoods.[10] Therefore, an analysis of the demography of subscribers may be required, and you may need to pay for having some homes connected to the cable.

An alternative would be to direct TV programming to receivers in facilities acquired expressly for the project. In addition to existing public schools, which might be used in the evening, day care centers, public housing auditoriums, or other community centers could be used. Such an approach will also make other modes of instruction more feasible. A third approach would be to direct all programming to the home and to supply other instruction or counseling at less frequent intervals in special facilities, as is done in the Open University and the Chicago TV College. This combines the advantages of both home- and classroom-based instruction.

In scheduling TV transmissions for a home-based audience, you will naturally take account of the nature of the audience. The home-*bound* will probably find that daytime programs fit their needs best, but job holders wanting to further their personal development must either rise earlier than usual or watch in the evening during prime time, when entertainment broadcasting competes for attention and channels. The Open University is likely to run into trouble when more than the four foundation courses are to be transmitted during prime time.[11] Here is where cable TV, with its promise of many channels, can solve a difficult problem.

[10] Park, *Prospects for Cable.*
[11] Maclure, "England's Open University Revolution."

VI. DEVELOPMENT AND OPERATION OF THE PROJECT

The planning activities described in the preceding chapter will lay the foundation for the development and operation phases of the project. Most of the steps in the development and operation phases displayed on Table 2, p. 200, are self-explanatory. There are a number of points, however, where additional suggestions may be helpful, because it is impossible to foresee and plan the development and operation of the project in complete detail.

Since most development activities are very similar to activities that will be carried out routinely during the operation of the project, the staff required will be similar to the operational staff. It is wise to involve those who will use the project in appropriate development activities, such as formative evaluation of pilot programs and preparation of utilization materials.

The development and operation activities that will be discussed here fall into three areas:

- The project
- Relations with independent groups
- The supporting organization

The Project

Activities important here are

- Preparation (or selection) of televised presentations
- Preparation of utilization activities
- Connection of schools and classrooms to the cable TV system and other program sources (wiring the project)
- Selection of personnel
- Preservice and in-service training

PREPARATION OF TELEVISED PRESENTATIONS

If you plan to have a significant portion of the televised presentations produced locally, some should be prepared before the project goes into operation to ensure that you will be ready to start project operation on schedule. Pilot programs will also be useful in preservice training and formative evaluation. The number of presentations that should be prepared beforehand will depend on how much time you have for development, how long it takes to produce a presentation, and whether you expect to have enough resources to keep pace with the demand for presentations during project operation. To answer these questions, refer to your schedule of televised presentations and plans for access to production facilities and equipment and estimate the total calendar time required to produce a televised presentation. To pin down the latter, you must estimate the degree of polish and sophistication desired for the televised presentations.

You will also need to consider that some presentational sequences require long lead times: They may use time-consuming techniques (such as animation), or must be produced at a distant spot, or require complex setups of materials or equipment, or have to be made at a certain time of year.

The lead time required for local productions will also depend on whether and to what extent you plan to preview and revise presentations before they are televised to the target audience. Preview and revision activities associated with the formative evaluation are discussed in Chap. 7; here we are concerned with activities that are part of the project monitoring mechanisms. The time required for such monitoring should be included as part of the production time for all programs. Most small projects lack the resources for this kind of monitoring.

SELECTION OF AVAILABLE MATERIALS FOR TELEVISED PRESENTATIONS

The same considerations apply to timing the selection and screening of available films and videotapes. We have discussed some information that will help guide the selection of such materials on p. 213. During screening, materials should be rated for[1]

- Clarity of purpose
- Clarity of presentation
- Appropriate rate of development
- Repetition
- Active participation by the learner

You should be able to tell, either beforehand or from viewing the presentation, what its general purpose is; that is, whether it is intended as supplementary or basic instruction, and how well it accomplishes its general goal. If the presentation is intended to instruct, can learning objectives be identified? Are there suggestions on

---

[1] Some of these characteristics and the subsequent discussion have been taken from C. R. Carpenter, "Instructional Film Research."

how to gather reliable and valid evidence on the learning that should occur? (Ideally, you would like testing materials to be supplied along with the presentation, as well as to have empirical evidence of its effectiveness with audiences similar to your target audience. Unfortunately, few presentations are accompanied by this kind of evidence and supporting material.)

Of course, you will expect the visual and auditory transmissions to be clear. Beyond this, however, clarity of exposition of facts, concepts, and theories is particularly important. Clarity may depend on simplicity in pictorial composition and on the amount of verbal commentary as well as its level of difficulty. The overall organization of the presentation should also be clearly developed and easily followed.

The rate of development of the presentation should be consistent with the level of difficulty of the material for the target audience. It should also take account of other tasks (such as taking notes or answering questions) that students are expected to accomplish during the presentation.

As for repetition, Carpenter says

> Repetition patterns of review and reiteration are strong variables that, with high frequency, produce significant effects in learning. Varied repetition is more effective for most cognitive learning tasks than merely using simple repetition. Varied repetition may aid in the avoidance of monotony and fatigue and in the processes of transfer and generalization of learning and skills acquisition.[2]

Although the research is ambiguous on the contribution of active student response to learning, according to Carpenter,

> the implications are strong that practice, participation, and involvement when properly managed in film production of many types increase the possibilities of learning, or they may be actual integral processes of the learning itself.[3]

Of course, the above are qualities that would be desirable in locally produced television presentations as well as in those that are screened from available sources.

PREPARATION OF UTILIZATION ACTIVITIES

Whatever means you have planned for activities that will integrate the televised presentations with the rest of the curriculum, you should begin their development along with the development of the televised presentations. Mismatches among availability dates for different portions of the project will, in effect, delay its full implementation and may undo much of the preliminary work in enlisting teacher and student support by creating disaffection with the project as a whole. Bear in mind that to be effective utilization activities must form bridges between the project and students, classroom teachers, and curriculum. Although the technical problems involved may be less than those associated with the televised presentations, the

[2] Ibid.
[3] Ibid.

coordination problems may be more severe. Hence, lead times for planning these activities and preparation of the needed materials can easily be equal to those required for obtaining the televised presentations. In addition, because utilization activities should be tailored specifically to your school system, it is unlikely that you will find that just what you want to use has already been prepared to accompany materials that are already available.

In the discussion of planning activities, we pointed out the desirability of monitoring project operations and listed the data and information that ought to be gathered (pp. 219–20). As development continues, the amount of resources devoted to monitoring will increase. The most effective channels for monitoring a project that reaches relatively few schools will probably be informal and may better develop as the project develops. If the project emanates from a point distant from many schools, however, you should formalize the monitoring function to ensure that the squeakiest wheels are oiled and that good ideas are not buried at the school or classroom level. These activities will be a permanent part of the project, in contrast to the formative and summative evaluations. Also, they should be conducted by school personnel; there is little to be gained by using independent evaluators in this role.

Unless the information gathered for project monitoring contributes directly to the collector's primary task (as data on student progress contribute to the teacher's instruction), you will need to take special steps to assure that the data collected is complete and accurate. Therefore, it may sometimes be wise for the project director or other key staff to make unannounced spot-checks of daily records. It is also important that what the monitor learns during his visits be relayed to the project director. A useful device here is the contact report, a printed form that the monitor fills out whenever he deals with someone in the school system and learns about some matter that should be brought to the project director's attention. Trip reports and debriefing meetings can also be used.

Finally, as a source of data there is no substitute for direct observations, conversations with teachers and other project participants, discussions with parents and students, and similar informal contacts.

WIRING THE PROJECT

There are several essential activities to be carried out during the development phase: connecting schools to the cable system, installing television sets in classrooms and auditoriums, installing intraschool wiring, and installing student response systems, if any. School administrators will have the responsibility for

- Designating the participating schools and classrooms.
- Monitoring the installation of the system for quality control.

In the choice of schools and classrooms for project participation you will be guided by several considerations:

- The size of the target audience in the given schools and classrooms.
- Support of principals and teachers.
- Accessibility to the cable system.

Ideally, the selection of schools and classrooms should reflect the objectives of the program. In addition, the target audience reached in each school should be large enough to justify the expense of installation. Support of principals and teachers is also desirable; if potential participants in the project are hostile, they can throw unexpected roadblocks into the selection process. Conversely, people eager to be part of the project may press for the wiring-in of a school that does not meet your criteria for the target audience. Finally, schools that are very remote from the cable system may be too expensive to include or may even be impossible to wire in because of technical limitations. Be sure to allow plenty of time for the selection process and for securing the cooperation of the principals and teachers chosen.

Check the performance of the cable system as it is being installed to make sure that the standards for reception are being met as agreed upon. In the past some school systems have not been provided with the quality they contracted for; it is well to catch these deficiencies early.

Wiring the project does not only involve the installation of the cable system, however. You should set up ways to provide teachers and students with all of the necessary materials. Neglect of mundane logistic matters such as delivery of study guides to the classroom teacher can nullify the best plans for classroom use. If this is to be an out-of-school project, materials may be mailed to students or delivered in other ways, such as by telephone or radio. Arranging for packaging, addressing, and mailing, for production of radio programs, for facilities for receiving telephone queries or conducting student conferences via telephone may all be part of wiring the project.

Once the project is in operation, preventive maintenance of the cable system, including the television receivers, and monitoring of associated means of delivering program materials should be routine. Whatever the scope of the project, you should make explicit arrangements for such matters and should set aside resources to ensure that they are carried out.

SELECTION OF PROJECT PERSONNEL

Teachers can be the key element in the project. Criteria for teacher selection might include: evidence of interest in and support for the project; past experience with similar students, subjects, and approaches; and evidence of ability to adjust to new situations. If teachers are to play an important role in the project, they must have the training, materials, and equipment required, and receive them on time.

For the host schools, the project may engender new scheduling requirements and special requests for facilities and services. Be sure to include the principal in planning, or he may view the project unsympathetically—as merely a source of added work. Devote special attention to keeping him in the decision loop and making sure that his interests are understood and respected.

PRESERVICE AND IN-SERVICE TRAINING

In the press of getting a project started, it is often difficult to provide good preservice training to the staff. In-service training may also be slighted because of other demands on staff time. This neglect is understandable, but it is unwise. Preservice training can serve two functions: preparing the staff to run the project and providing valuable direction for subsequent project development.

In training, you will show the staff how to use the televised presentations and other components of the project. This means that you will need to give trainees a good sample of the materials and presentations they will use. Such training can most usefully take place after a few complete learning units have been produced but long enough before the project starts (at least several weeks) to allow the trainees to become familiar with the units.

You should arrange well in advance for the facilities, materials, equipment, and students that will be needed for the training sessions. Positive inducements, such as overtime payments, should be offered to trainees for attending training sessions. Trainees should include all staff who will be directly involved with the project—those responsible for televised presentations, classroom teachers and paraprofessionals, and those responsible for production of utilization materials. Train these people as a team, using the sample learning units as vehicles for working out roles and responsibilities. Many teachers want students to be involved in preservice training sessions so that the teachers can practice handling real problems.

More in-service training will be needed if the teacher's role is very different from that in the regular classroom. Teachers and other staff should have frequent opportunities to discuss and resolve the problems that are bound to arise and to share useful and rewarding experiences.

Relationships with Independent Groups

As your project develops, you are apt to find that the set of independent groups that supports your efforts will change and that more fruitful liaisons will emerge with the passage of time. Remain flexible in this regard and be ready to shift liaison efforts as need be.

An innovative project is likely to lead to many requests for visits. Teachers and administrators from other school systems, reporters, researchers, and cable TV operators may want to visit studios and classrooms to observe the project in operation. To avoid undue disruption, establish special visiting days and set up procedures to control who can visit under what conditions and when. However, teachers who are not in the project should be encouraged to observe it in action and be given the time and the means to do so. If visits must be limited, give teachers top priority. Visiting teachers should also be encouraged to attend in-service training or consulting sessions to learn about the project—both its good and bad features.

Supporting Organization

During the development phase, personnel slots will be filled, and the organization to support the project will begin to grow. During this process, questions will arise as to who should do what and who has the right to decide what. If the project is small, decisions on these matters can be made informally. If many people are involved, reliance on informal agreements may result in the unequal division of labor, ruffled feelings, and slipped schedules. To keep lines of authority and responsibility clear, the project manager should be readily available to his staff to settle disputes. The monitoring processes will also help alert him to problems.

VII. EVALUATION

Introduction

If your project does not depart radically from regular classroom practice, information and data supplied from project monitoring mechanisms may satisfy your needs for evaluation. If, however, your project is innovative, will make significant use of funds from outside sources, or is an experiment or demonstration in some other sense, you should plan to include evaluation as part of the initial development and operating phases. Evaluation is commonly regarded as a postproject function, but it should be planned when other project operations are planned to ensure that decision-relevant information is obtained. Without substantial evaluation planning, there is a high probability of poorly selected or designed measuring instruments, untested students, poor test conditions, or incomplete or irrelevant criteria.

Planning the Evaluation

The four steps in developing an evaluation plan are

- Determining the functions of the evaluation.
- Selecting the evaluator.
- Selecting evaluation measures.
- Defining project success in operational terms.

DETERMINING THE PURPOSE OF THE EVALUATION

The evaluation may have one or all of the following purposes:

- To assist in project improvement, or *formative* evaluation.
- To help sponsors decide about the future of the project, or *summative* evaluation.

- To provide information that will assist other education agencies in implementing a similar project, or *transference* evaluation.

Summative evaluation is evaluation in the traditional sense. But most traditional evaluations have been skimpy; in particular, evaluators have tended to ignore the resource requirements of a project, a consideration obviously vital to sponsors. We shall discuss below the kinds of data and information that should be included in a summative evaluation.

Formative evaluation has only recently been recognized as a valuable function.[1] It is a conscious attempt to step back from the project during its development and initial operation to assess its strengths and weaknesses. You can use such information to improve the project as it is formulated, thereby increasing its chances of success. Formative evaluation is obviously closely related to project monitoring, but you should view the latter as a permanent part of project operation, which is usually not true of formative evaluation.

In the past transference evaluation has not been recognized as an evaluative function. The current emphasis on providing model projects focuses attention on the need for a description of the model that is sufficiently complete that another education agency can adapt the model to its own use.[2] This means that enough information should be provided that the user will also have confidence in the outcome. A sponsor viewing your project as a model for others will be particularly concerned with transference evaluation.

SELECTING THE EVALUATOR

The choice of an evaluator depends on his assigned functions. If the summative evaluation requires him only to measure changes in student learning or behavior brought about by the project, there are many universities, nonprofit groups, education agencies, and business firms that have the required resources and skills. Surveys of the audience for *Sesame Street,* for example, were conducted by the Educational Testing Service, a TV audience rating service, students working on advanced degrees, Children's Television Workshop's own utilization department, and various research bureaus. Alternatively, you may want to have the test and measurement staff of your own school system provide the required measures. It is better to use an independent evaluator if the sponsor wants a patently unbiased assessment. For example, faculty at the University of Southern California conducted the summative evaluation for Anaheim.

If the evaluator is to perform a complete summative evaluation, a new set of qualifications must be met because data will be needed on resource requirements as well as on project effectiveness. You will want an organization that can work well with the officials involved and that has the required skills in resource and education analysis. You may wish to draw on management consultants to gather some of these

[1] Rapp, *Evaluation as Feedback.*

[2] See, for example, Bretz, *Three Models for Home-Based Instructional Systems;* and Haggart, *Program Cost Analysis.*

data. If so, you should clarify the division of responsibility among the various people assessing project outcomes.

Formative evaluation levies the largest demands for skills and resources because it requires frequent visits to observe, interview, and collect data. The evaluator must establish rapport with project personnel and win their trust, while maintaining objectivity about the project's results. Because it may be difficult to find people in your own schools who have the needed skills and at the same time do not have an emotional commitment to some facet of the project, it might be wise to have an outside agency do the formative evaluation. For example, the formative evaluation for *Sesame Street* was performed by the Educational Testing Service.

Data and information provided by both the formative and summative evaluations will be useful for transference evaluation. Additional information will be needed, however. Other education institutions that are thinking of following your model will want to know how the project was implemented, what pitfalls were encountered, and what aspects were crucial to project success. This means that running accounts of project planning, development, and initial operation will be useful, as will assessments by the staff of keys to success and of misdirected efforts. Therefore, project staff should contribute to the transference evaluation from the outset. Perhaps you should give this responsibility to one staff member within each component of the supporting organization. Responsibility for final collation, analysis, and presentation of transference information logically rests with whoever does the summative evaluation.

SELECTING EVALUATION MEASURES

To provide a solid basis for evaluation, whatever its role or the objectives of the project may be, you will need data and information in most of the following areas:

Project Outputs

- Number reached in target audience.
- Numbers in target audience achieving personal goals (primarily for out-of-school projects).
- Educational services provided.
- Cognitive growth attributable to project.
- Affective change attributable to project.
- Psychomotor growth attributable to project (not applicable to most projects).
- Attitudes toward the project.

Project Operation

- Performance of the cable TV system and classroom reception system.
- Performance of the program production or acquisition components.
- Operation of the utilization component.
- Operation of the supporting organization.

Project Inputs

- Incremental requirements for dollars.
- Requirements for specialized staff.
- Requirements for other scarce resources.

The precise measures you choose for the outputs of the project and the importance of the measures relative to each other will depend on the project goals. And, as many of the items in the list suggest, the summative evaluation requires baseline data and information, that is, data and information describing the situation in each area that would have existed without the project. Baseline data may be historical, but too frequently the data needed will not have been gathered in preceding years. Thus, it is often useful to designate groups of students as "controls" or "comparisons."[3] Another approach is to use the treatment group as its own comparison; again, it is unlikely that all of the data needed to describe the treatment group will have been gathered.

Number Reached in Target Audience

If your project involves courses for which students may register, registrations will give you a lower bound on the size of the audience. We know that many more people view courses than are registered for them. For example, the Chicago TV College has data that indicate that there are 250,000 regular viewers each semester, and perhaps 500,000 frequent, but not daily, viewers of each semester's courses. This compares with a total enrollment (both for credit and not-for-credit) in 1967-68 of only 11,718.[4]

If formal enrollment is not part of your project, special audience surveys will be needed to determine the size of the viewing audience. Even if the project is a required part of your school curriculum, actual classroom use may be overestimated if it is equated with broadcast time, which is currently the practice in most school systems.[5] The most informative index of use of televised presentations would be the numbers of students viewing each presentation. Such information might be gathered by direct observation of samples of presentations in samples of classrooms. Responses to questionnaires may be biased for a number of reasons. The evaluators used several methods to determine the size and character of the viewing audience of *Sesame Street:* face-to-face interviews, telephone interviews, survey questionnaires mailed to schools with follow-up on schools not returning the questionnaires, and television rating services.

Numbers in Target Audience Achieving Personal Goals

Some projects, particularly out-of-school projects for adults, contribute directly to personal goals, such as obtaining a certificate of equivalency of a high school

---

[3] It should be borne in mind that it is usually very difficult, if not impossible, to structure a bona fide scientific experiment in the real world of a school system. For example, strictly comparable control groups rarely exist, experimental data are contaminated by personal biases, information is missing, and so on.

[4] Schmidbauer, *Multimedia System in Adult Education.*

[5] Wagner et al., *A Study of Systemic Resistances to Utilization of ITV in Public School Systems.*

diploma, receiving credit for a course in higher education, receiving a degree in higher education, entering a training program or getting a job as a result of a career counseling program, or simply attaining a new skill such as gourmet cooking or a new appreciation of cultural activities. Some data, such as course credits, will be gathered routinely. The evaluator will have to make special efforts to gather others.

Educational Services Provided

This item represents the educational content of the project—academic subject matter, guidance services, information on school activities for the community, teacher training, or student productions for television. Such information will be a part of the description of the project, and little effort will be needed to obtain it. It is simply an integral part of the information needed for any evaluation and can play a key role if services are provided that would not otherwise have been available.

Cognitive Growth Attributable to Project

If the project content falls into standard academic or professional fields, you can probably use standardized tests as measuring instruments, as was done in Hagerstown and Anaheim. However, the usual standardized tests were not designed to measure, for individual students, the short-term effectiveness of instruction, and consequently, the use of such tests for this purpose can present statistical and practical problems. This is why the Hagerstown and Anaheim summative evaluation designs, each of which spanned several years, are particularly exemplary.[6]

Criterion-referenced or learning mastery tests are an alternative to standardized tests. On learning mastery tests, student mastery is judged by criteria related directly to the subject matter rather than by the student's standing with respect to his peers nationwide. A review of experience in past projects, however, now shows that there are severe problems with such tests:[7]

- The needed tests probably do not exist; if they have to be devised after the project is started, testing schedules will probably not be met.
- Few instruments have been field-tested for reliability.
- Criterion-referenced tests are hard to administer: Some require special training of test administrators, and others are tied to individualized curricula forcing the evaluator to use sampling techniques for students and objectives to keep required resources within reasonable bounds.
- Unless the tests are given on a pretest and posttest basis, you will not know whether the students had already mastered the objectives before entering the project.

[6] Wade, "Hagerstown," and Anaheim City School District, "Summary of Instructional Television Evaluation."

[7] The problems were dramatically illustrated in the OEO performance contracting experiment. Twenty-five percent of the contractors' pay was to be based on criterion-referenced examinations, which OEO called interim performance objective tests (IPOs). The large number of tests and test items required by the individualized curriculum and the changes in curricula during the school year made it unrealistic for contractors to submit tests in advance to the evaluator for analysis and approval. The result was, according to OEO, that some of the tests were too easy and others failed to measure what contractors had taught. The OEO report concluded that "the IPOs appear to have been virtually useless for evaluation purposes and to have had questionable value for payment purposes." *An Experiment in Performance Contracting.*

If your resources permit you to mount an extensive evaluation effort, however, tests tailored to your curriculum may be devised and administered by the evaluator. In Anaheim, for example, the evaluator constructed special tests in science, social sciences, and Spanish,[8] for *Sesame Street,* the evaluator devised tests cued to the *Sesame Street* curriculum and administered them to each child, individually, at each time of testing.[9]

Affective Change Attributable to the Project

Evaluators often use survey questionnaires to assess attitude changes, but these generally have low validity and sometimes low reliability. Trained psychometrists can assess attitudes fairly reliably, however, through structured interviews. Ideally, such interviews should be validated by direct observation of student behavior in appropriate situations. Another approach is to infer affective changes from behavioral data such as vandalism rates, dropout rates, police actions, disciplinary problems, or attendance rates. The task is easier at the high school level, where such variables can sometimes be regarded as reflecting student attitudes toward school. At the elementary level, attendance seems to be the best index. Referrals to the principal for discipline are also a relevant measure but they depend strongly on the tolerance of teachers toward deviant behavior.[10]

Attitudes toward the Project

If attitudes are assessed at all, it has usually been the attitudes of project participants (primarily students and teachers) toward the project itself and toward the use of television for teaching in particular. Teacher and student attitudes in these areas can vary from hostile to highly supportive. Supportive attitudes, particularly on the part of teachers in an in-school project, probably enhance project effectiveness and certainly make the project director's job easier. A student's attitude toward televised teaching may, however, have little correlation with his academic performance.[11]

If you plan to expand your project in future years, opinions of teachers not directly involved in the project may be important. You may wish to include all teachers in the school system in any opinion surveys you conduct. Reactions of administrators, cable TV operators, leaders of other involved groups, and parents are also relevant, and it is wise to ensure that all factions are heard, not only the most vocal. The influence of the project on parental attitudes may turn out to be the most important measure of success for some projects. Recorded test gains may be no more than modest, but cold facts and figures do not carry the force of those

[8] Anaheim City School District, *Summary of Instructional Television Evaluation.*

[9] Reeves, *The First Year of Sesame Street.*

[10] There has been little stress on bringing about affective change through educational television programs. But affective results can be measured and can provide a good basis for project decisions. The 1970-71 performance contracted project for dropout prevention in Texarkana is a case in point. Despite low gains on standardized achievement tests, the evaluator recommended, and the school system adopted, a policy of continuing the project without contractor participation primarily because of a sharp reduction in the dropout rate. Hall et al., *A Guide to Educational Performance Contracting.*

[11] Lee, *Test Pattern.*

parents who believe that educational quality has been improved. Participation by parents in activities designed to involve them in the project is a valid indicator of genuine interest.

The least expensive way to assess attitudes toward the project is by means of a paper-and-pencil questionnaire (or an orally administered questionnaire for young children). Such a questionnaire was administered to the faculty of the Scarborough TV College by Professor D. S. Abbey, one of the faculty members. Its main finding was that the faculty agreed with the statement "Although its value has not been fully assessed, we should be willing to give television a try." However, as Lee describes,

> Ironically, Professor Abbey departed from the statistically demonstrable data to note that in his *conversations* with Scarborough professors, "many instructors" negated this statement. He concluded that agreement with the statement in the questionnaire might be "possible acquiescence" to the "expected" response from a university staff member.[12]

There is really no substitute for face-to-face interviews conducted by people not directly involved in the project. At the least, a paper-and-pencil questionnaire should be anonymous to encourage honest responses. Even so, respondents may succumb to their natural desire to please and seem more positive than they actually are.

Project Operation

Information concerning the interior workings of the project (sometimes termed *process evaluation*) will be needed for the formative evaluation, and the history of project development will be needed for the transference evaluation. Adequate monitoring mechanisms will ensure that such information is gathered routinely.

Any evaluation must, of course, rest on a solid understanding of what "the project" actually is. Initially, it will be developmental, altering as it goes along to adjust to unforeseen problems and to take advantage of unexpected opportunities. Although the end result may well be a more effective project, in its ultimate successful mutation it may seem like a different species from the project described initially. The evaluator should note major changes in the instructional process, in the use of resources, in the groups of teachers or students involved, or in the project administration. Otherwise, "the project" that is continued or expanded in the future may be an imperfect model of what actually went on.

Project Inputs

The resources you should consider in planning the project have been listed on Tables 6 and 7, pp. 223–24. The summative and transference evaluation should display the incremental resources that actually were used, along with an estimate of additional resources required to continue the project or to expand or change it in some other way if change is being contemplated. The formative evaluation will keep

[12] Ibid.

track of resource consumption to catch instances of obvious overconsumption or unexpected savings.

In addition to dollars, you should note the project's requirements for specially trained personnel, specially designed equipment, or other resources that are likely to be in short supply.

DEFINING PROJECT SUCCESS

To define the success of your project, rank a combination of measures in the areas just discussed. The ranking you choose will directly reflect the general objectives of the project. Let's recall the four general uses to which a cable TV education project might be put:

- Increased physical accessibility of education.
- Provision of additional educational services.
- Improved quality of education.
- Decreased unit cost of education.

Increased Physical Accessibility of Education

If your primary objective is to make education more accessible to a particular target audience, the most important measure will be the number reached in the target audience. You may also be concerned with the numbers achieving personal goals and probably with cognitive or psychomotor growth attributable to the project. Another significant measure will be an estimate of the resources that would have been required for the target audience to obtain the benefits of the project in some other way. If these are large compared with the resources required by the project, you may have the strongest selling point you can find.

Provision of Additional Educational Services

This objective is similar to the first in the sense that if the incremental resources required by the project are less than those that would have been required to provide the services in some other way, you may count the project a success. You will also need to show that the services were effectively provided, which will require measures of growth by project participants or assessments of attitudes toward the project.

Improved Quality of Education

What is important here are the measures of growth and attitudes toward the project. You expect the project to require more resources than would be needed without it; you and the sponsor must judge whether the improvement in quality is worth the cost.

Decreased Unit Cost of Education

Once again, the issue of resources required is a prime factor. You will be in the position of having to compare the incremental resources required by the project with the cost of providing the same services in the traditional way. You must also be able to argue, however, that the quality of education has not suffered. Hence, some measures of growth will still be needed.

THE NEED FOR FLEXIBILITY

Unless the steps discussed above have been taken in formulating the evaluation plan, it is unlikely that the evaluation will yield the information you need. There is also, however, the danger that the plan can become cast in concrete. We have already noted that you can expect your project to change its complexion during the development and initial implementation phases, as detailed information on its strengths and weaknesses becomes available. Thus, the plan must be flexible enough to accommodate to change. Otherwise, you may end up evaluating a project that doesn't exist or even allowing the evaluation to hamper the successful implementation of the project. This is a real danger, as the experience during the first year of the Texarkana performance contracting program demonstrated.[13]

Evaluation during Project Development

As the project is being developed, the following evaluative activities may be undertaken:

Summative evaluation: Collect baseline data, redefine measures.

Formative evaluation: Direct pilot efforts and assess their results, collect data on resources being consumed, assess attitudes toward project, counsel project staff, monitor project operation, monitor cable TV installation and operation.

Transference evaluation: Collect baseline data, collect data on resources being consumed, assess attitudes toward project, chronicle project planning and development phases.

ACTIVITIES FOR SUMMATIVE EVALUATION

During this activity you will probably discover that some of the data and information you want are deficient on a number of counts. One possible difficulty is that the data are invalid measures of the characteristic you are trying to capture. For

---

[13] Carpenter, Chalfant, and Hall, *Case Studies in Educational Performance Contracting: 3*, pp. 21-23.

example, absences due to dislike of school in the previous year may have been camouflaged by absences due to a flu epidemic. Even more serious is when the data are so incomplete or inaccurate they are essentially useless. Sometimes you can find this out only be interviewing school personnel at length. In either event, you should redefine your measures so that the baseline data (or qualitative information) you need can be collected before thay become lost in the operation of the project.

ACTIVITIES FOR FORMATIVE EVALUATION

If you have sufficient resources, use pilot TV programs as vehicles for formative evaluation during project development. Pilots should be miniature models of essential features of the full project; usually they are learning units complete with the televised presentation, accompanying utilization activities and some of the before-and-after measures that are planned for the summative evaluation. Pilots tell project staff not only how effective the educational portions of the project are but also how well the various components of the project work together. They also serve as vehicles for preservice training.

Careful formative evaluation of pilot programs during the development phase was an outstanding characteristic of *Sesame Street*, as described below:

> Individual segments of the show were produced in the spring of 1969, and tested by the research staff with target audience samples, in order to determine the probable educational effectiveness (comprehension and achievement) of program content and format.
> Five pilot shows were produced and tested in two cities by the research staff during the summer of 1969 for educational effectiveness and program appeal.[14] Recommendations for program refinements, based on the results of this study, were transmitted to the production staff, and testing techniques were also refined as a result of this work.[15]

No other projects in the past have, to our knowledge, incorporated a formative evaluation as comprehensive as that for *Sesame Street*. Many, however, have used rehearsals and pretesting on a small scale to give producers and teachers practice and to iron out various kinks.

The other activities listed above for formative evaluation during the development phase (collect data on resources being consumed, assess attitudes toward program, counsel program staff) are self-explanatory.

ACTIVITIES FOR TRANSFERENCE EVALUATION

We shall discuss only the chronicle of planning and development activities, which should tell others wishing to adapt the project to their own needs what steps were taken, where snags occurred and why, and what were the most successful

---

[14] The research staff developed special techniques for measuring program appeal during this work.
[15] Bretz, *Three Models for Home-Based Instructional Systems*. Note that the formative evaluation contributed to the summative evaluation.

approaches. This chronicle should include the staff's hindsight on how the project could have been implemented more efficiently and some estimate of the time and resources saved thereby. Since the project will presumably be innovative (if this kind of evaluation is justified), false steps and wasted motion during the initial phases are to be expected. Therefore, the evaluator should make it clear that he is not pointing a finger at mismanagement but rather transmitting valuable lessons to others who may want to use the project as a model.

Evaluation after Project Implementation

After the project has been in operation for a time, resources devoted to formative evaluation will diminish and project monitoring mechanisms will take over. Data and information will continue to be gathered for the summative and transference evaluations, however, until the first phase of project operation has been completed. At that time, sponsors, project directors, school system administrators, and others will decide whether to continue the project in its present form, to change it, to drop it, or to implement a variant of it elsewhere.

The evaluation should provide a description of the development of the project in order to answer such questions as

- What in fact was "the project"?
- How long was the project actually in effect?
- Did start-up problems reduce project effectiveness?
- What were the major obstacles to implementation?
- How did attitudes toward the project change as it was implemented?
- What project features seemed to generate the most enthusiasm or to work the most smoothly?

A major function of the evaluation is to gather quantitative data on project effects. Analysis of these data will reveal whether the project had effects and how extensive they were. As noted previously, the use of control groups is very helpful at this point. In most situations, the data warrant only relatively simple analysis. Elaborate analyses of incomplete or inaccurate data, or sophisticated comparisons of data that are not really comparable, are a waste of time and misleading.

The significance of effects on student learning is obvious. An analysis of test items will also provide useful data for project assessment by illuminating its weak and strong parts. This will point the way to whatever changes in content or emphasis are desirable.

The evaluation should also assess major nonquantifiable effects, such as changes in degree of acceptance or support for the project on the part of teachers, administrators, parents, community groups, teacher unions, and, of course, the students themselves. Changes in the classroom environment are also important.

There may be other intended or unintended outcomes from the project, which the evaluator should describe. Unanticipated side effects may be key considerations in decisions about the future of the project. For example, the Experiment in Community Communication in Newburgh, New York, had its greatest effect on the young people who operated the studio. For many of them, it was their first chance to accomplish something on their own, and it encouraged some to move into other community services.[16]

Choice of Alternatives

The evaluation should provide direction for the design and comparison of alternative project configurations for the future. One such alternative, of course, is for the school system to revert to its prior practices—an alternative that, if nothing else, can provide a baseline for judging the incremental effects of other configurations.

Another alternative is to adopt an entirely different approach. A radical change would only be of interest if the project just completed were patently unsuccessful or unacceptable to the school system, if the educational need remained as pressing as ever, or if some potential alternative had been tried in another school system with outstanding results. The disadvantage, of course, is that the school system would have to go through the implementation process all over again.

There will also be a set of alternatives based on the project just completed. Changes might be made in several areas:

- **Content.** Should some subject matter be deleted? Added? Are changes in emphasis needed?
- **Students.** Should the project be extended to more students? Should a student population with different characteristics be included?
- **Staffing.** Should the mix of students, teachers, and paraprofessionals be changed? Should the staff be acquired, trained, and paid in the same ways?
- **Management.** Should the project continue to be managed in the same way?
- **Evaluation.** Should the evaluation be conducted as before?

Although many alternatives could be generated by all possible answers to these questions, probably only a few will need to be considered.

After assessing the educational process and outcomes, you should analyze the resource requirements of the alternatives to estimate the resource impacts of different projects on the district's inventories, personnel, and cash flow. This will provide input to financial planning and resource management on one hand, and information for determining the scope of the alternatives on the other. Such a planning exercise should have a time horizon that spans the initial phase-in period and any anticipated major modifications. It should also include at least the first year in which costs and operations are expected to level off.

You should collect resource information on the candidate projects, incorporate this information into a design for the planned implementation of the projects, and then arrive at schedules of resource requirements. Delete resources that are available without additional cost to the school system to obtain a list of the incremental resources that the system must acquire.

We suggest a seven-step analysis:

[16] Price and Wicklein, *Cable Television.*

1. Summarize the resource requirements of the candidate projects.
2. Derive resource factors.
3. Describe the scope of the planned implementation of the projects.
4. Estimate future resource needs for the planned projects from steps 1 and 2.
5. Subtract, from district costs, those resources that will be available without cost.
6. Summarize the incremental impacts on resource inventories and staffing levels on a year-by-year basis.
7. Display summary project costs on a year-by-year basis.[17]

Given a set of alternative projects accompanied by descriptions of their resource requirements and their effectiveness, you can compare them on the basis of their relative returns. Such a comparison is often termed a cost-effectiveness analysis. However, in the educational context, answers can rarely be more than rough approximations, for it is seldom possible to compute accurately a cost-effectiveness ratio derived by statistical methods from an abundant data sample.

There are two ways to conduct a cost-effectiveness analysis. One is to specify the level of effectiveness desired and then compare the costs of alternative means to achieve it. The other is to specify an acceptable cost level and compare the relative effectiveness of projects whose costs cluster around that level. In analyzing instructional projects, the fixed-budget approach is probably the most helpful. That is, you should say "We have $100,000 to spend on 350 students. Which project will give us the greatest effectiveness?"

In evaluating a project and deciding on its future, these two salient points should be kept in mind:

1. Examine a broad range of project outcomes—favorable and unfavorable—in addition to cognitive test scores; and
2. Compare the cost and effectiveness of the possible implementations of the project relative to each other and to the situation without them.

[17] For further guidance, see Haggart, *Program Cost Analysis.*

Appendix A

COMPARISON OF THE COSTS OF ITFS AND CABLE SYSTEMS FOR SUPPLYING EDUCATIONAL/INSTRUCTIONAL TV CHANNELS TO SCHOOLS
by Carl Pilnick [1]

The intent of this appendix is to compare (under a set of simplifying assumptions) the costs of distributing educational or instructional TV program material via an *ITFS broadcast system* or a *coaxial cable network*.

For the comparison, a school district with the following characteristics is assumed:

1. One central studio and distribution center.
2. A total of 10 schools with an average of 500 students per school. Eight schools are located within a 6-mile radius of the studio and two within 25 miles.
3. Four channels of program material (the maximum normally granted under an ITFS license).
4. Two-way communication (not mandatory initially).

ITFS Configuration

Figure 5 illustrates the assumed location of the 10 schools. Ideally, the transmitting antenna(s) for an ITFS system should be located, as shown in Fig. 5, as equidistant as possible from the receiving sites. Whether this is feasible, however, depends on the topography of the area, since the transmitting antennas should also be as high as possible to achieve maximum line-of-sight coverage. Thus, if a centralized location is relatively flat, a high transmitting tower (with its attendant cost) would be necessary. If a naturally elevated location such as a mountain top is accessible, its height may make it a more suitable location for transmission even if it is not equidistant from the station receiving antennas.

---

[1] President, Telecommunications Management Corporation.

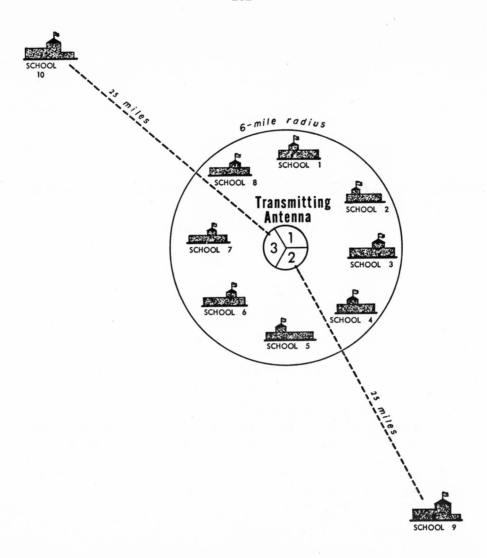

Fig. 5—Configuration of school district

For this cost comparison, a centralized transmitter is assumed, which may require either a tall free-standing tower or a smaller tower on the top of a convenient building.

Three separate transmitting antennas are assumed. The total transmitter power for an ITFS station is limited by the FCC to 10 watts. It is estimated that the bulk of this power, perhaps 6 watts, will feed antenna 1, which will provide an omnidirectional radiation pattern designed to reach schools 1 to 8.

About 2 watts each will feed to transmitting antennas 2 and 3, which would be highly directional parabolic or dish antennas beamed at schools 9 and 10, respectively.

At receiving sites 1 to 8, relatively inexpensive receiving antennas would be adequate, while schools 9 and 10 would require narrowly directive receiving antennas, possibly 8- to 10-foot parabolas, to match the narrow-beam characteristics of the transmitted signals.

Figure 6 illustrates the block diagram of the major elements of the ITFS system. Whether live, on videotape, or converted to video from film, program material emanates from the studio/distribution center. If the transmitter is not located at the studio site, which is probable, the video signals are linked via a coaxial cable.

The transmitter is shown as a 4-channel multiplex unit, capable of transmitting up to 4 TV-format programs simultaneously, using the assigned ITFS broadcast frequencies in the 2500-2686 MHz band. The same signals are fed to each of the three transmitting antennas.

At each school, a receiving antenna is required. The received signals are then connected, through cable, to a down-converter, which changes the broadcast frequency of the 4 channels from the 2500 MHz band to standard VHF frequencies used for TV, so that the programs will be compatible with ordinary TV receivers. In the Los Angeles area, for example, the frequencies for channels 3, 6, 8, and 10 could be used since no over-the-air broadcasts are assigned to these.

The number, type, and cost of the TV receivers in the schools, and the in-school cabling from a central distribution point to each classroom, are assumed to be the same for either ITFS or cable, so that no cost differential exists for this portion of the system.

A typical response or inquiry system requiring reverse communication is shown by the dashed lines of Fig. 6. At each TV set, a push-button actuated microphone could be located with the audio signal modulating an FM transmitter that operates in the talk-back section of the ITFS band (125 KHz within the upper 4MHz (2686-2690) of the ITFS spectrum).

This FM signal is transmitted from a school-top antenna to the central tower location, down-converted and delivered to a speaker, or perhaps a data terminal at the studio/distribution center. With a similar installation at each school, any student could at any time enter an audio or alphanumeric response into the system.

ITFS Equipment Costs

Table 8 presents an estimate of the costs of an ITFS system as configured by Figs. 5 and 6.

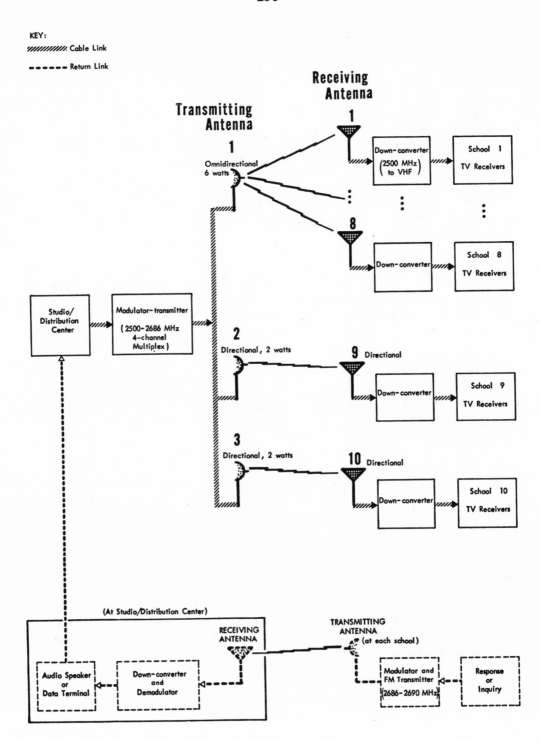

Fig. 6—ITFS distribution system

Table 8

ESTIMATED COSTS, ITFS SYSTEM

10 schools, 500 students per school

(in dollars)

Component	One-way Communication	Two-way Communication
A. Items identical for ITFS and cable systems		
1. Studio/distribution center, color, 4-channel capability	160,000	
2. TV receivers (and stands), color, assume 20 per school @ $400	80,000	
3. In-school cabling, from central distribution point to classrooms @ $4,000 per school	40,000	
4. Coaxial cable link from studio to transmitter (or cable system head-end)	5,000	
B. ITFS--one-way communication		
1. Transmitting tower	25,000	
2. Transmitter, 4-channel multiplex, and 3 transmitting antennas	60,000	
3. Receiving antennas and down-converters 8 @ $1,500 2 @ $8,000	28,000	
C. ITFS--two-way communication (audio only)		
1. Push-button actuated microphones (1 per TV receiver @ $50)		10,000
2. FM modulator-transmitter 10 @ $5,000		50,000
3. FM transmitting antennas 8 @ $500 2 @ $4,000		12,000
4. FM receiving antenna and down-converter		5,000
5. Audio distribution and switching equipment, central studio		1,000

Cost Category	One-way Communication	Added Cost: Two-way Communication
Total equipment cost	398,000	78,000
Equipment cost per year (8-year amortization period)	49,750	9,750
Operating cost per year (including rental of buildings and facilities, exclusive of programming)	100,000	---
Maintenance cost per year (10% of equipment cost)	39,800	7,800
Total cost per year	189,550	17,550
Cost per student per year (@ 5,000 students)	37.91	3.51[a]

[a]Total = $41.42 per student per year.

Group A components represent those items that would be required in essentially the same form whether an ITFS or a cable system were used. Group B includes the major ITFS items necessary for one-way communication, that is, sending TV program material from the distribution center to the 10 schools. Group C includes the talk-back components required for interactive, two-way communication (audio/visual downstream, audio only upstream).

The amortization period for the capital equipment has been chosen somewhat arbitrarily at 8 years. Arguments can be made for any interval between 5 and 10 years.

The operating costs ($100,000 per year) are typical of actual operating ITFS systems, such as at Stanford University, but can vary considerably depending on the amount of student and faculty labor available to assist in system operation. Because this cost alone represents $20 per year per student, any major variation would be significant.

Cable System Configuration

For an all-cable distribution system, certain further assumptions must be made. Either a CATV system may be available to provide spare channel capacity for educational programming, or a dedicated cable system is necessary for school use only. Both cases are considered below.

Case 1: CATV System Available

In the major urban areas CATV systems are now required by the FCC to provide at least one educational channel free (for the first 5 years); additional channels may be available for *lease*. To achieve a 4-channel system, it is assumed therefore that 3 channels must be leased by the school district.

To estimate the cost of leasing 3 channels, a dual-cable CATV system without converters is assumed, providing a total of 24 channels via a cable selector switch. These 3 channels represent 3/24 or 1/8 of the normal system's capacity, and on a pro rata basis may also be said to represent 1/8 of the CATV system cost.

This approach probably represents a cost at the high extreme, because the CATV operator depends on subscriber revenue from retransmission of commercial TV for his primary return, and any spare channel capacity would provide *incremental* rather than *proportional* revenue increases. Thus, although we realize that this is a conservative estimate, we can still use it for comparison.

To estimate the total construction cost of the CATV system, if the school district contains 5000 students, the total population in the area might be 30,000 people, with perhaps 10,000 homes. If 50-percent subscriber penetration is realized, there would be 5000 CATV subscribers, and at a rule of thumb of $100 per subscriber, the CATV system cost would be *$500,000*.

One-eighth of this cost, for the 3 leased channels, would therefore be $62,500, which is the pro rata school district construction cost.

Figure 7 illustrates the CATV system distribution method, while Table 9 summarizes the various costs.

It is assumed that the two distant schools, 9 and 10, can best be served by providing separate microwave links to each rather than to attempt extending cable trunks such a distance.

Referring to Table 9, Group A items are identical with Table 8. In the Group B category are the pro rata construction costs of 3 channels of the CATV system, as estimated above, and the 2 microwave links required to service schools 9 and 10.

For Group C items, it is assumed that the CATV system includes two-way communications capability within its basic cost. Because schools 9 and 10 are not connected to the cable, however, separate FM talk-back facilities are necessary in their case.

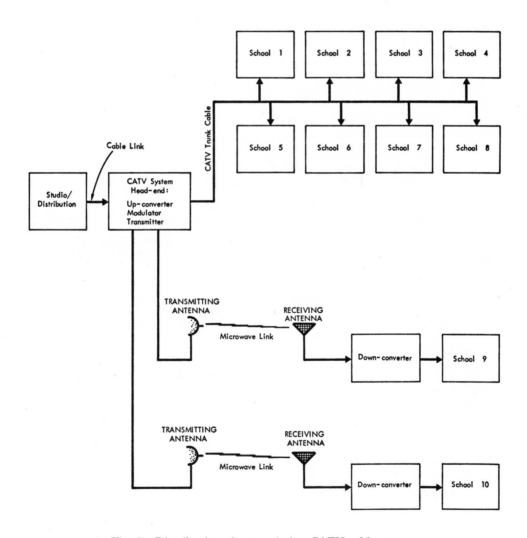

Fig. 7—Distribution via an existing CATV cable system

Table 9

ESTIMATED COSTS, EXISTING CATV SYSTEM

(in dollars)

Component	One-way Communication	Two-way Communication
A. Items identical for ITFS and cable systems		
1-4. Same as Table 8	285,000	
B. Cable system components--one-way communication		
1. Pro rata cost of 3 channels of 24-channel CATV system	62,500	
2. Microwave links to schools 9 and 10, up-converters, modulators, transmitters and antennas, receiving antennas, down-converters	40,000	
C. Cable system--two-way communication		
1. Push-button actuated microphones (same as Table 8)		10,000
2. Schools 9 and 10 FM, transmitter-receiver systems 2 modulator-transmitters @ $5,000 2 transmitting antennas @ $4,000 1 receiving antenna and down-converter @ $5,000		23,000
3. Audio distribution and switching equipment, central studio		1,000

Cost Category	One-way Communication	Added Cost: Two-way Communication
Total equipment cost	387,500	34,000
Equipment cost per year (8-year amortization period)	48,438	4,250
Operating cost per year (exclusive of programming)	60,000	---
Maintenance cost per year (5% of equipment cost)	19,375	1,700
Total cost per year	127,813	5,950
Cost per student per year (@ 5,000 students)	25.56	1.19[a]

[a]Total = $26.75 per student per year.

A reduction in both operating and maintenance costs is shown in Table 9, because the CATV system operators will share most of this burden. The operating cost given ($60,000), which accounts for nearly half of the total annual cost, is a proration of the cost to operate the entire cable TV system. It could easily be half as much as estimated here, which would make cable TV an even more attractive option.

Case 2: No CATV System Available

If no commercial CATV system is or will be available in the school district, we can consider a totally dedicated cable system.

Referring to Fig. 5, if we assume an average of 3 miles distance between each school (schools 1 to 8 only), about 25 to 30 miles of coaxial cable are required to link a head-end to schools 1 to 8. Normally, a CATV system costs $4000-$5000 per cable mile to construct, but in this case since only the schools and not a multitude of homes must be connected, the costs will be much lower (assumed at $2500 per mile). Thus, 30 trunk miles will cost $75,000.

Separate microwave links to schools 9 and 10 are still assumed.

Table 10 shows the equivalent costs for a dedicated cable system. The operating costs are assumed equal to that of an ITFS system, and maintenance is estimated at 10 percent of equipment cost per year since the complete cable network is now the maintenance responsibility of the school district.

Summary

The examples used as the basis for estimates provide only a guide to comparative cost evaluation, and it is stressed that any particular situation must be examined in detail with respect to objectives and trade-off considerations.

With this caution, Tables 8 to 10 indicate that if an existing CATV system provides coverage for much of the school district area, a considerable cost saving can be effected (in the vicinity of 40 to 50 percent) by "piggybacking" the educational program distribution on the CATV facilities. This is true even if a substantial channel leasing charge is applied.

If a fully dedicated system is necessary, Tables 8 and 10 indicate approximately equivalent costs (for 4 channels) for either an ITFS or a special purpose cable system. Again, this equivalence may be quite misleading unless other factors are considered. As one example, the cable network is capable of providing many more channels at relatively small incremental cost, while the ITFS system, under FCC rule, is limited to 4.

Table 10

ESTIMATED COSTS, DEDICATED CABLE SYSTEM

(in dollars)

Component	One-way Communication	Two-way Communication
A. Items identical for ITFS, cable, and dedicated cable systems		
1-4. Same as Tables 8 and 9	285,000	
B. Dedicated cable system--one-way communication		
1. Cost of special cable system	75,000	
2. Same as Table 9	40,000	
C. Dedicated cable system--two-way communication		
1. Same as Table 9		10,000
2. Same as Table 9		23,000
3. Same as Table 9		1,000

Cost Category	One-way Communication	Added Cost: Two-way Communication
Total equipment cost	400,000	34,000
Equipment cost per year (8-year amortization period)	50,000	4,250
Operating costs per year (exclusive of programming)	100,000	----
Maintenance cost per year (10% of equipment cost)	40,000	3,400
Total cost per year	190,000	7,650
Cost per student per year (@ 5,000 students)	38.00	1.53[a]

[a]Total = $39.53 per student per year.

Appendix B

MAJOR SOURCES OF ITV PROGRAMMING

There are two main sources of ITV programming, each of which will be described briefly. These sources can also provide you with information on producers of ITV programming, both education agencies and commercial firms.

Great Plains National (GPN) Instructional Television Library

Tracy Clement
P.O. Box 80669
Lincoln, Nebraska
Telephone: (402) 467-2502

GPN is in the business of collecting and disseminating instructional television programming, primarily in the form of videotape. It also handles film and is producing programming in videocassette format. Information on TV courses provided in its catalog includes subject area, title, number of lessons, lesson length in minutes, whether color or black and white, the producer, and an abstract of the content. In addition to curricular materials, courses are offered for training teachers in TV or film use.

Almost all videotaped courses are offered for lease only. The user provides his own videotape on which the course is freshly duplicated from the master (although he may request GPN to provide the tape for an extra charge). The user pays for duplication, handling fees, and royalties. Films may be either leased or purchased.

Courses may be previewed free of charge by means of a preview kit, comprising one or more representative lessons along with a copy of related printed matter.

For the lease of a single 20-minute, videotaped lesson, GPN charges a flat rate of $50. A lesson for videocassette costs $96.20, which includes the cost of the cassette.

GPN has undertaken the implementation of a set of systematic procedures for designing instruction for television when it is used primarily to teach. These procedures have been applied in eight test sites, with favorable results. GPN's approach to developing instructional television was recognized by both the U.S. Office of Education and the Association for Educational Communications and Technology of the National Education Association as "the only major national commitment to the application of systems analysis and common sense to the design of instruction for television."[1]

National Instructional Television Center (NIT)

Bill Perrin
Box A
Bloomington, Indiana 47401
Telephone: (812) 339-2203

NIT's primary concern is to produce and disseminate courses to fit the curricular needs of particular school systems. Although its major activity is to produce new courses, NIT acquires or adapts locally produced ITV materials as well.

As with GPN, courses may be leased or purchased, and free preview services are available. Rental fees depend on the number of students to be served. For a 20-minute, videotaped lesson, the lease cost is between $36.50 and $84.

Previews should be requested at least three weeks in advance of the desired preview date; orders for rental of courses must reach NIT at least six weeks before the starting date of the series. Lessons must be used in sequence in consecutive school weeks at the rate of one lesson per week. After use, the lesson must be shipped to the next user. This allows NIT to bicycle lessons from one user to the next, thereby holding costs down. Other arrangements are possible, and regional libraries have been established to expedite matters; nevertheless, the operation appears fairly unwieldy.

Other Sources

The list of educational TV stations that produce programming for local education agencies is too unstable and too long to be included here. For example, the first tenth of the GPN catalog for 1972 contained programming produced under the sponsorship of:

Central Virginia ETV Corp, Richmond, Va., WCVE-TV
North Central Council for School Television,
    Fargo, N.D., KFME-TV
Bradley University TV Center, Peoria, Ill.,
    WIRL-TV
Greater Washington TV Assn., Inc., Washington,
    D.C., WETA-TV
University of Calgary, Department of
    Communications Media

[1] "Lincoln Leadership Conference Supports GPN ITV Design Approach," *Great Plains National ITV Library Newsletter*, University of Nebraska, Lincoln, Nebraska, February 1972.

San Diego Area Instructional Television
   Authority, KEBS-TV
Denver Public Schools, KRMA-TV
University of Illinois, WILL-TV
Nebraska Council for ETV, KUON-TV
New York City Public Schools, WNYE-TV

Probably the best way to find out if producers of instructional programming are near you is to contact GPN or NIT.

There are other organizations that offer assistance in the selection of ITV programming. None, however, provide general search or rating services. Indexing and retrieval services are provided by the National Information Center for Educational Media (NICEM) at the University of Southern California, University Park, Los Angeles, California 90007, telephone: (213) 825-1323. NICEM has indexed over 200,000 16-mm films, 35-mm filmstrips, audio tapes, videotapes, 8-mm motion cartridges, overhead transparencies, and audio records. Indexes include information on subject area, audience level, color or black and white, sound reproduction requirements (where applicable), duration or number of frames, producer, and distributor. The index to producers and distributors lists over 10,000. NICEM will also catalog an institution's holdings.

A number of federal agencies produce materials (primarily 16-mm films) that can be rented or bought from the National Audiovisual Center, Washington, D.C. 20409, telephone: (301) 440-7753. Most of these materials are directed to adult audiences and deal with matters of vocational, consumer, or public interest.

GLOSSARY

Amplifier: An electrical device that amplifies the voltage, current, or power of an electrical signal.

Cable television, cable TV, CATV: A system for distributing audiovisual information via coaxial cable. Includes its signal receiving, amplifying, and controlling equipment and signal origination equipment. The system is used under franchise granted by a governmental body.

Channel: A frequency band providing a single path for transmitting electrical signals, usually in distinction from other parallel paths.

Closed-circuit television, closed-circuit TV, CCTV: A system for distributing audiovisual information via coaxial cable. Includes its signal receiving, amplifying, and controlling equipment and signal origination equipment. The system is under control of the user (see Dedicated system).

Coaxial cable: The most commonly used means of signal transportation for cable TV, consisting of a cylindrical outer conductor surrounding a central conductor.

Common carrier: Microwave service available at published tariff rates approved by the FCC.

Computer assisted instruction, CAI: The use of a computer in a tutorial mode. The computer adapts the presentation of instructional material on the basis of inputs from the student.

Dedicated system: A transmission system operated entirely for a single user or a small group of users. The user determines content, scheduling, points of reception, and other matters having to do with program transmissions.

Digital data: Data and information that can be represented by a set of discrete items such as the integers.

Drop: The connection between the cable TV system and the subscriber's TV set.

Educational television, ETV: All noncommercial television intended for general use. The term includes public affairs, cultural, educational, entertainment, or other programming. The medium of broadcast television is generally used. The term includes ITV and PTV.

Film chain: A device for showing sound, motion pictures on TV.

Headend: The electronic equipment located at the start of a cable system, usually including antennas, preamplifiers, frequency converters, demodulators, modulators, and related equipment. May also include the antenna tower and building housing the above.

Independent random access: Access to an item of instructional materials (including audiovisual materials) at random times and independent of the use of the item by another person.

In-school project: A project for transmitting television programming to the facilities of an education agency.

Instructional Television Fixed Service, ITFS: A service operated by an education organization to transmit audiovisual information to one or a few fixed receiving locations. Transmission is provided at frequencies higher than those used for broadcast television.

Instructional television, ITV: Television programming that has as its purpose the production, origination, and distribution of instructional content for people to learn. It is closely related to the work of organized formal education agencies.

Local production: Production of television programming by a cable TV operator, closed-circuit TV operator, ITFS operator, or other local agency, rather than by television networks or national agencies.

Microwave: Applies to transmission at frequencies well above the normal television frequencies.

Narrow-band: A range of frequencies less than 3000 cycles from lowest to highest.

Out-of-school project: A project for transmitting television programming to the home or other facilities that are not part of an education agency.

Penetration: The percent of total homes that have one or more television sets connected to a cable TV system, or being connected to a cable TV system, at the time of a survey.

Plasma panel: A computer terminal device for visual display. It consists of bubbles of gas that can be made to glow on and off by electric current.

Preferential treatment: The provision of cable system facilities or services to a subscriber free or at a lower cost than that charged the general subscriber.

Public television, PTV: ETV programming intended for the general community, as opposed to ITV.

Random access: Access to an item of instructional materials (including audiovisual materials) at random, as distinct from scheduled, times. Access may depend on whether the item is already in use by another person, however.

Trunk: The main cable in a cable system.

Two-way capability: An ability to transmit signals in both directions. Return signals may be digital, audio, or audiovisual.

Utilization: The use of televised programming for its intended (usually educational or instructional) purpose.

Utilization activities: Activities that abet utilization.

Video: Relating to or used in the transmission, reception, or recording of audiovisual information. Usually considered to be signals in the range of normal television frequencies.

Videotape: Magnetic tape carrying audiovisual information.

BIBLIOGRAPHY

Anaheim City School District, Anaheim, California, "Summary of Instructional Television Evaluation."

Baer, Walter S., *Cable Television: A Handbook for Decisionmaking*, Crane, Russak & Co., New York, 1974.

——, *Interactive Television: Prospects for Two-Way Services on Cable*, The Rand Corporation, R-888-MF, November 1971.

——, and R. E. Park, "Financial Projections for the Dayton Metropolitan Area," Paper 2 in Johnson et al., *Cable Communications in the Dayton Miami Valley: Basic Report*, The Rand Corporation, R-943-KF/FF, January 1972.

Belt, Forest H., "Education in Kentucky—By Television," *BM/E (Broadcast Management and Engineering)*, October 1971.

Bloom, B. S. (ed.), *Taxonomy of Educational Objectives. Handbook I: Cognitive Domain*, David McKay Company, Inc., New York, 1956.

Bogatz, Gerry Ann, and Samuel Ball, *The Second Year of Sesame Street: A Continuing Evaluation*, Educational Testing Service, Princeton, New Jersey, November 1971.

Bretz, R., "Closed-Circuit ITV Logistics—Comparing Hagerstown, Anaheim, and Santa Ana," *Journal of the NAEB*, July-August 1965.

——, "The Potential Uses of Cable in Education and Training," Paper 7 in Johnson et al., *Cable Communications in the Dayton Miami Valley: Basic Report*, The Rand Corporation, R-943-KF/FF, January 1972.

——, *The Selection of Appropriate Communication Media for Instruction: A Guide for Designers of Air Force Technical Training Programs*, The Rand Corporation, R-601-PR, February 1971.

——, *A Taxonomy of Communication Media*, Educational Technology Publishers, Englewood Cliffs, New Jersey, 1970.

——, *Three Models for Home-Based Instructional Systems Using Television*, The Rand Corporation, R-1089-USOE/MF, October 1972.

——, *Will My Visual Be Visible?*, The Rand Corporation, P-4919, October 1972.

The Carnegie Commission on Educational Television, *Public Television: A Program for Action*, Bantam Books, New York, January 1967.

The Carnegie Commission on Higher Education, *The Fourth Revolution—Instructional Technology in Higher Education*, McGraw-Hill Book Company, New York, June 1972.

Carpenter, C. R., "Instructional Film Research—A Brief Review," *British Journal of Educational Technology,* Vol. 2, No. 3, October 1971.

Carpenter, P., A. W. Chalfant, and G. R. Hall, *Case Studies in Educational Performance Contracting: 3. Texarkana, Arkansas, Liberty-Eylau, Texas,* The Rand Corporation, R-900/3-HEW, December 1971.

——, and S. A. Haggart, *Analysis of Educational Programs within a Program Budgeting System,* The Rand Corporation, P-4195, September 1969.

——, and G. R. Hall, *Case Studies in Educational Performance Contracting: Conclusions and Implications,* The Rand Corporation, R-900/1-HEW, December 1971.

——, and M. L. Rapp, *Testing in Innovative Programs,* The Rand Corporation, P-4787, March 1972.

——, et al., *Analyzing the Use of Technology To Upgrade Education in a Developing Country,* The Rand Corporation, RM-6179-RC, March 1970.

Chu, G. C., and W. Schramm, *Learning from Television: What the Research Says,* Stanford University, Stanford, California, 1967.

Connell, D. D., "How Sesame Street Raised and Spent Eight Million Dollars," *Dividend,* Graduate School of Business Administration, University of Michigan, Winter 1971.

Cooney, J. G., *The Potential Uses of Television in Preschool Education,* New York, October 1966.

——, *Television for Preschool Children: A Proposal,* February 19, 1968.

Educational Product Report, No. 31, Instructional Television Fixed Service, Educational Products Information Exchange Institute, New York, 1971.

Educational Product Report, No. 36, Dial Access Systems and Alternatives, Educational Products Information Exchange Institute, New York, 1971.

Efron, E., "Peter Chelkowski, Ph.D., I Love You," *TV Guide,* November 21, 1970.

Erickson, C. G., and H. M. Chausow, *Chicago's TV College, Final Report of a Three Year Experiment,* Chicago City Junior College, Chicago, August 1960.

Etzioni, Amitai, "Human Beings Are Not Very Easy To Change After All," *The Saturday Review,* Vol. 55, No. 23, June 3, 1972.

Fact Book—Television Instruction, Indiana University Foundation, National Instructional Television Center, Bloomington, Indiana, 1972.

Farquhar, J. A., et al., *Applications of Advanced Technology to Undergraduate Medical Education,* The Rand Corporation, RM-6180-NLM, April 1970.

Federal Communications Commission, "Cable Television Report and Order," *Federal Register,* Vol. 37, No. 30, Part II, February 12, 1972.

——, "Memorandum Opinion and Order on Reconsideration of the Television Report and Order," *Federal Register,* Vol. 37, No. 136, Part II, July 14, 1972.

Feldman, N. E., *Cable Television: Opportunities and Problems in Local Program Origination,* The Rand Corporation, R-570-FF, September 1970.

——, "System Designs for the Dayton Metropolitan Areas," Paper 1 in Johnson et al., *Cable Communications in the Dayton Miami Valley: Basic Report,* The Rand Corporation, R-943-KF/FF, January 1972.

Haggart, S. A., *Program Cost Analysis in Educational Planning,* The Rand Corporation, P-4744, December 1971.

——, G. C. Sumner, and J. R. Harsh, *A Guide to Educational Performance Contracting: Technical Appendix,* The Rand Corporation, R-955/2-HEW, March 1972.

——, et al., *Program Budgeting for School District Planning,* Educational Technology Publications, Englewood Cliffs, New Jersey, 1972.

Hall, G. R., et al., *A Guide to Educational Performance Contracting,* The Rand Corporation, R-955/1-HEW, March 1972.

Hammond, A. L., "Computer-Assisted Instruction: Two Major Demonstrations," *Science,* Vol. 176, No. 4039, June 9, 1972.

Hill, Roger W., Jr., "Educational Considerations of CATV Cablecasting and Telecommunications," *Educational/Instructional Broadcasting,* Vol. 2, No. 9, November 1969.

Johnson, L. L., *Cable Television and Higher Education: Two Contrasting Examples,* The Rand Corporation, R-828-MF, September 1971.

——, *The Future of Cable Television: Some Problems of Federal Regulation,* The Rand Corporation, RM-6199-FF, January 1970.

——, et al., *Cable Communications in the Dayton Miami Valley: Basic Report,* The Rand Corporation, R-943-KF/FF, January 1972.

Jones, G. R., *The Jones Dictionary of CATV Terminology,* Johnson Publishing Co., Boulder, Colorado, 1971.

Kenney, B. L., and F. W. Norwood, "CATV: Visual Library Service," *American Libraries,* July-August 1971, pp. 723-726.

Klaus, D. J., *Instructional Innovation and Individualization,* American Institutes for Research, Pittsburgh, Pennsylvania, 1969.

Kratochvil, Daniel W., *Sesame Street, Developed by Children's Television Workshop,* Product Development Report No. 10, American Institutes for Research in the Behavioral Sciences, Palo Alto, California, December 1971.

Lee, J. A., *Test Pattern,* University of Toronto Press, Toronto, 1971.

Levien, R. E., *The Emerging Technology: Instructional Use of the Computer in Higher Education,* McGraw-Hill Book Company, New York, 1972.

Maclure, S., "England's Open University Revolution at Milton Keynes," *Change,* Vol. 3, No. 3, March-April 1971.

Martin-Vegue, C. A., Jr., et al., "Technical and Economic Factors in University Instructional Television Systems," *Proceedings of the IEEE,* Vol. 59, No. 6, June 1971.

McCarty, H. R., *Proposal for Development of Electronic Communication System for the San Diego Region,* Department of Education, San Diego County, California, January 17, 1972.

McCombs, M., "Chicago's Television College," *New Educational Media in Action: Case Studies for Planners,* Vol. 2, United Nations Educational, Scientific and Cultural Organization, Holland-Breumelhof N.V., Amsterdam, 1967.

McLaughlin, G. H., et al., *Educational Television on Demand,* Occasional Papers No. 11, The Ontario Institute for Studies in Education, Toronto, 1972.

Molenda, M. H., "The Educational Implications of Cable Television (CATV) and Video Cassettes: An Annotated Bibliography," *Audiovisual Instruction,* Vol. 17, No. 4, April 1972, pp. 42-59.

National Education Association, *Cable Television: Franchise Provisions for Schools, Instruction and Professional Development,* NEA, Washington, D.C., February 1973.

——, *A Survey of Instructional Closed-Circuit Television, 1967,* NEA, Washington, D.C., 1967.

1972 Recorded Instruction for Television, University of Nebraska, Great Plains National Instructional Television Library, Lincoln, Nebraska, 1972.

Noel, E. S., and G. M. Helmke, *Instructional Television in California,* California State Department of Education, Sacramento, California, 1968.

Park, R. E., *Prospects for Cable in the 100 Largest Television Markets,* The Rand Corporation, R-875-MF, October 1971.

Price, M., and J. Wicklein, *Cable Television: A Guide for Citizen Action,* Pilgrim Press, Philadelphia, Pennsylvania, 1972.

Quick, John, and Herbert Wolf, *Small-Studio Video Tape Production,* Addison-Wesley Publishing Co., Reading, Massachusetts, 1972.

Rapp, M. L., *Evaluation as Feedback in the Program Development Cycle,* The Rand Corporation, P-4066, April 1969.

Reeves, B. F., *The First Year of Sesame Street: The Formative Research,* Children's Television Workshop, New York, December 1970.

Reuben, Gabriel H., "Using Cable Television To Involve Parents," *Educational Television,* III, January 1970.

Robinson, W. R., *Analysis of Data from the IRTV Overload Study,* Bell Telephone Research, Ltd., Ottawa, Canada, May 25, 1971.

Rockman, S., *One Week of Educational Television, No. 6, March 9-15, 1970,* National Instructional Television Center, Bloomington, Indiana, 1971.

Rubinstein, E. A., et al., *Television and Social Behavior, Vol. IV, Television in Day-to-Day Life: Patterns of Use,* U.S. Government Printing Office, Washington, D.C., 1972.

Samuels, B., *The First Year of Sesame Street: A Summary of Audience Surveys,* Children's Television Workshop, New York, December 1970.

Schmidbauer, M. (Project Manager), *Multimedia System in Adult Education,* Internationales Zentralinstitut für das Jugend- und Bildungsfernsehen, Munich, 1971.

Schramm, Wilbur, et al., *The New Media: Memo to Educational Planners,* United Nations Educational, Scientific and Cultural Organization, Holland-Breumelhof N.V., Amsterdam, 1967.

Shanks, Robert E., "The Anaheim Approach to Closed-circuit Television," *A Guide to Instructional Television,* ed. Robert M. Diamond, McGraw-Hill Book Company, New York, 1964.

"Teacherless Classes," *Nation's Schools,* Vol. 89, No. 6, June 1972.

Television Factbook, Services Volume, No. 41, Television Digest, Inc., Washington, D.C., 1971-1972 edition.

Tickton, S. G. (ed.), *To Improve Learning: An Evaluation of Instructional Technology,* 2 vols., R. R. Bowker Company, New York, 1970, 1971.

U.S. Department of Health, Education, and Welfare, Office of Education, National Center for Educational Statistics, *Bulletin,* No. 7, February 9, 1971.

U.S. Department of Labor, Office of Economic Opportunity, *An Experiment in Performance Contracting: Summary of Preliminary Results,* OEO Pamphlet 3400-5, February 1972.

U.S. Government Films: A Catalog of Motion Pictures and Filmstrips for Rent and Sale by the National Audiovisual Center, General Services Administration, National Archives and Records Service, Washington, D.C., Publ. 72-17, 1971 Supplement.

Wade, S., "Hagerstown: A Pioneer in Closed-Circuit Televised Instruction," *New Educational Media in Action: Case Studies for Planners,* Vol. 1, UNESCO, International Institute for Education Planning, Amsterdam, 1967.

Wagner, R. V., et al., *A Study of Systemic Resistances to Utilization of ITV in Public School Systems, Volume II, Case Studies,* American University, Washington, D.C., February 1969 (ED 030 012).

Wigren, H. E., "The NEA's Position on Cable Television," *Schools and Cable Television,* Division of Educational Technology, National Education Association, Washington, D.C., 1971.

Contributors

POLLY CARPENTER-HUFFMAN is a research analyst in the Management Sciences Department of The Rand Corporation. Prior to joining Rand in 1955, she taught at Ohio State University and in the public schools of Florida. Her current research interests center on the instructional process, measures of effectiveness in education, planning educational programs, and the design of instructional systems. She received her B.A. in education and M.A. in mathematics from the Ohio State University. She is the author of numerous reports in the domestic and international education fields.

RICHARD C. KLETTER is a consultant to the Communications Policy Program of The Rand Corporation, as well as to foundations and community organizations. He served as Director of the Portola Institute's Media Project in Palo Alto, California, and has written research reports on cable television, videocassettes, and other communications developments during his graduate studies at Stanford University's Institute for Communications Research. He holds a B.A. in political science from Colgate University and an M.A. in communications research from Stanford University.

ROBERT K. YIN is a research psychologist with The Rand Corporation and serves part time as Assistant Professor of Urban Studies and Planning at the Massachusetts Institute of Technology. His primary research interests are the delivery of municipal services to neighborhoods, neighborhood change and social indicators, the social and service impact of telecommunications, and the relationships between citizens and governments. He holds a B.A. in history from Harvard, an M.A. in government and public administration from George Washington University, and a Ph.D. in social psychology from MIT.

WALTER S. BAER directed the Rand cable television study and edited this series of volumes. His research interests include the development of two-way communications on cable systems, issues of media ownership, and the technical and economic prospects for new communications systems. He has been a communications consultant to the United Nations and to a number of major corporations. He also has directed the Aspen Workshop on Uses of the Cable sponsored by the Aspen Program on Communications & Society. Previously he served on the White House science advisory staff and as a White House Fellow in 1966-67. He received his B.S. in physics from the California Institute of Technology and a Ph.D. from the University of Wisconsin.

SELECTED RAND BOOKS

Bagdikian, Ben H. *The Information Machines: Their Impact on Men and the Media.* New York: Harper and Row, 1971.

Bretz, Rudy. *A Taxonomy of Communication Media.* Englewood Cliffs, N. J.: Educational Technology Publications, 1971.

Bruno, James E. (ed.). *Emerging Issues in Education: Policy Implications for the Schools.* Lexington, Mass.: D. C. Heath and Company, 1972.

Cohen, Bernard, and Jan M. Chaiken. *Police Background Characteristics and Performance.* Lexington, Mass.: D. C. Heath and Company, 1973.

Coleman, James S., and Nancy L. Karweit. *Information Systems and Performance Measures in Schools.* Englewood Cliffs, N. J.: Educational Technology Publications, 1972.

Dalkey, Norman C. (ed.). *Studies in the Quality of Life: Delphi and Decisionmaking.* Lexington, Mass.: D. C. Heath and Company, 1972.

DeSalvo, Joseph S. (ed.). *Perspectives on Regional Transportation Planning.* Lexington, Mass.: D. C. Heath and Company, 1973.

Downs, Anthony. *Inside Bureaucracy.* Boston, Mass.: Little, Brown and Company, 1967.

Fisher, Gene H. *Cost Considerations in Systems Analysis.* New York: American Elsevier Publishing Co., 1971.

Haggart, Sue A. (ed.). *Program Budgeting for School District Planning.* Englewood Cliffs, N. J.: Educational Technology Publications, 1972.

Harman, Alvin. *The International Computer Industry: Innovation and Comparative Advantage*. Cambridge, Mass.: Harvard University Press, 1971.

Levien, Roger E. (ed.). *The Emerging Technology: Instructional Uses of the Computer in Higher Education*. New York: McGraw-Hill Book Company, 1972.

Meyer, John R., Martin Wohl, and John F. Kain. *The Urban Transportation Problem*. Cambridge, Mass.: Harvard University Press, 1965.

Nelson, Richard R., Merton J. Peck, and Edward D. Kalachek. *Technology, Economic Growth and Public Policy*. Washington, D.C.: The Brookings Institution, 1967.

Novick, David (ed.). *Current Practice in Program Budgeting (PPBS): Analysis and Case Studies Covering Government and Business*. New York: Crane, Russak and Company, Inc., 1973.

Park, Rolla Edward. *The Role of Analysis in Regulatory Decisionmaking*. Lexington, Mass.: D. C. Heath and Company, 1973.

Pascal, Anthony H. (ed.). *Racial Discrimination in Economic Life*. Lexington, Mass.: D. C. Heath and Company, 1972.

Pascal, Anthony H. *Thinking about Cities: New Perspectives on Urban Problems*. Belmont, Calif.: Dickenson Publishing Company, 1970.

Quade, Edward S., and Wayne I. Boucher. *Systems Analysis and Policy Planning: Applications in Defense*. New York: American Elsevier Publishing Company, 1968.

Sharpe, William F. *The Economics of Computers*. New York: Columbia University Press, 1969.

Williams, John D. *The Compleat Strategyst: Being a Primer on the Theory of Games of Strategy*. New York: McGraw-Hill Book Company, 1954.